The Unofficial Guide to vMix®

BY PAUL RICHARDS

DEDICATION

To the Streaming Idiots and the amazing group of StreamGeeks who have
helped all along the way!

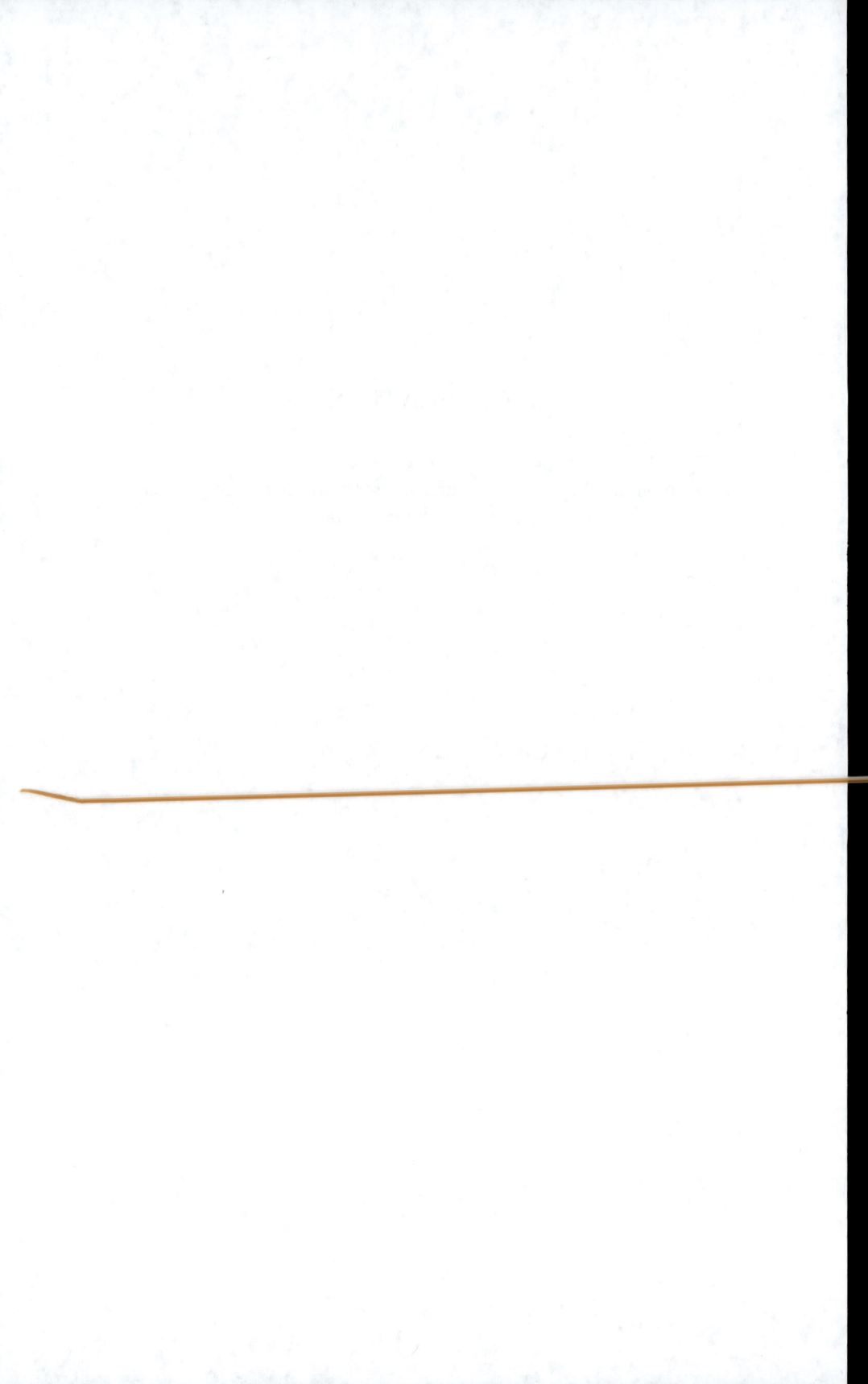

CONTENTS

ACKNOWLEDGMENTS

I would like to acknowledge my production team mates Michael Luttermoser, Tess Protesto, Julia Sherwin, Brian Mulcahy, and Lindsey Pope.

ONLINE COURSE

Consider taking your learning further with the optional online course for this book on Udemy here: https://www.udemy.com/course/vmix-live-streaming.

FACEBOOK GROUP

Consider joining the other StreamGeeks using vMix in Facebook group here: http://facebook.com/groups/streamgeeks.

1 WHAT IS VMIX?

vMix is live video production software that can turn a regular Windows computer into a professional video production studio. The software allows you to mix together video and audio sources into a production which can be recorded, streamed and connected to a number of popular video production workflows. The output of vMix can be set up in standard definition (SD), high definition (HD), and even 4K. All you need is a PC desktop or laptop with Windows 10 and a DirectX10 compatible graphics card.

More specific system requirements are available at https://www.vmix.com/software/supported-hardware.aspx.

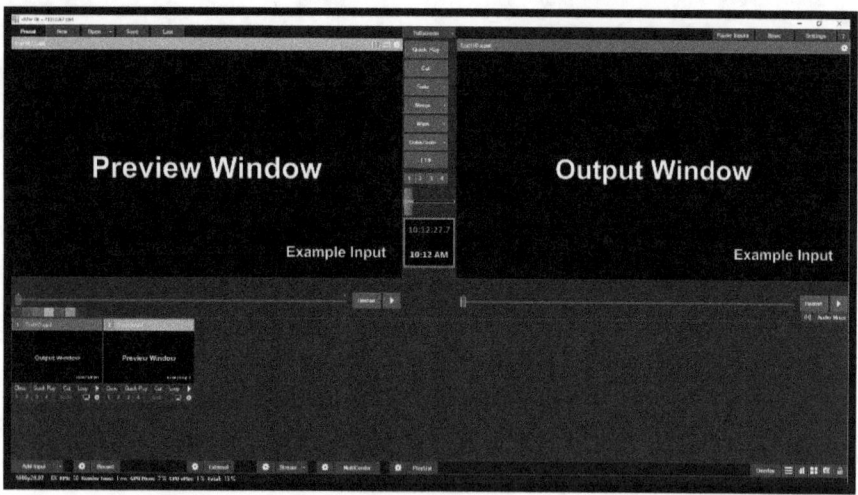

The layout of vMix is designed to create the look and feel of a professional broadcast studio with both preview and output windows set side by side during operation. Some users may feel a bit overwhelmed at first but will quickly find the interface both intuitive and powerful.

What Are You Going to Learn in This Book?

In this book, you will learn everything you need to know to set up and operate vMix like a pro. First, you will learn how to download the software, get set up and become familiar with the interface. Next, you will learn about your audio-visual mixing options and common workflows associated with vMix productions. Then, you will learn how to get all your sources connected.

vMix accepts inputs in multiple formats, including cameras, capture devices, NDI (Network Device Integration) sources, video files, and even more advanced input sources such as SRT and RTSP live video streams. You will then learn about all the features within the software you can use to mix sources and improve the look of your video. Next, you will be introduced to options for controlling and viewing your video production. Finally, you will see how easy it is to begin your own production for recording or live streaming. You can even use the virtual camera output from vMix to send audio and video into a webinar or video conference software.

How Does vMix Compare to Other Solutions in the Market?

There are several options for video production software, but vMix hits the sweet spot of features and value for many users and organizations. It has a unique pricing strategy that allows users with different budgets and needs to choose the best option. It comes in five editions Basic, Basic HD, HD, 4K, and Pro, ranging in price from free to $1200.

Many of the best core features are included in even the lowest-priced editions. The lowest-priced paid option ($60) offers HD resolution, three camera inputs, overlays, built-in animated titles, scoreboards, and tickers. All versions allow the user to record and send up to three simultaneous live streams. One of the greatest features of vMix is its ability to grow with you. New streamers can purchase a license for what they need at the time and quickly upgrade, adding more features without having to learn a new software environment.

For this reason, vMix is a great choice for many users because it isn't likely you will have to stop the learning process and switch to another software because of a technical limitation. vMix is used for simple productions and advanced solutions everyday all around the world. In fact, if you keep an eye out, you will find the world's top broadcast professionals using vMix in even the most mission critical situations.

2 GETTING STARTED

Part of what makes vMix great is that it is available in six editions to fit your exact needs and budget. If you are not sure what works best for you, you can try the full edition for free for 60 days. That way, you can see what features and options you need before you decide.

The six editions differ in the features that they unlock. With the free edition, you get a fully functional program with four inputs, two camera inputs, and a maximum resolution of 768 x 576 (SD). It also includes the ability to send three simultaneous live streams and record. You can take advantage of one video overlay channel and access built-in animated titles, scoreboards, and tickers.

The Basic HD edition adds an additional camera input and bumps resolution up 1080p. The next level (HD) virtually eliminates input limits with a total of 1000 inputs for cameras or any other source. It also adds vMix Call for one caller, which is great for bringing in remote guests to your production. It also offers a total of four overlay channels. The final two editions, 4K and Pro, increase the top resolution to 4K plus add more advanced features like dual recording, instant replay, and PTZ camera control.

Installing vMix

Whether you have already purchased a vMix license or want to try the free 60-day trial with full functionality, the process is the same:

Go to vMix.com

1. Click on the download button near the top of the screen.
2. Click the button to begin the download.
3. Run the installer.
4. Follow the installation prompts.
5. Enter your email to begin the free trial or enter your vMix registration key that you received after a purchase.

Once the software is installed and running, it will prompt you to select a preset. Do not worry if you aren't sure about this. You can change it later. vMix presets are used to save layouts of inputs. These presets can be saved and loaded as an easy way to switch and recall projects you have worked on in the past. You will want to pick the option that matches the resolution and frame rate of your cameras or other video sources. If you have installed the software on an older, lower-power computer, you may need to choose a lower resolution and frame rate to ensure smooth operation.

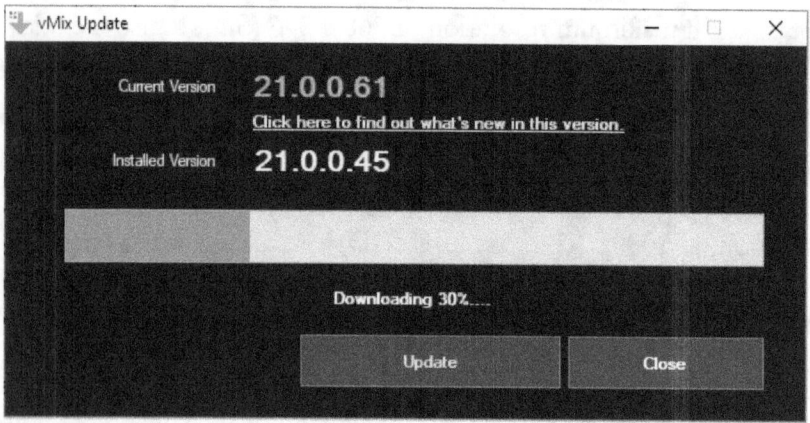

Updating vMix

One of the great things about vMix as a company, is their support for new features. The creators of vMix pay close attention to their community of users who actively post new ideas in the user forums. If there is enough support for a new feature idea, it is usually implemented in the software within a period of time. As new updates become available, you will see a little message at the bottom of vMix that says, "Update Available."

Each purchase of vMix includes 12 months (1 year) worth of free updates. After one year, you will need to purchase additional updates. You can check to see if you are eligible for free updates and purchase additional updates by logging into your vMix account at https://account.vmix.com/.

Finding Help

Once you are up and running, be sure to go back to vMix.com to get all the information and support you may need to get started. Clicking on the "support" button on the home page will give you access to the user guide and training videos. You can also search the knowledge base for answers to frequently asked questions and known issues, join the forums to share knowledge with other users, and even reach out to support if you can't find the information you need.

Now with the software installed and access to all the help you may need, it is time to dig in and get started with vMix.

vMix Interface

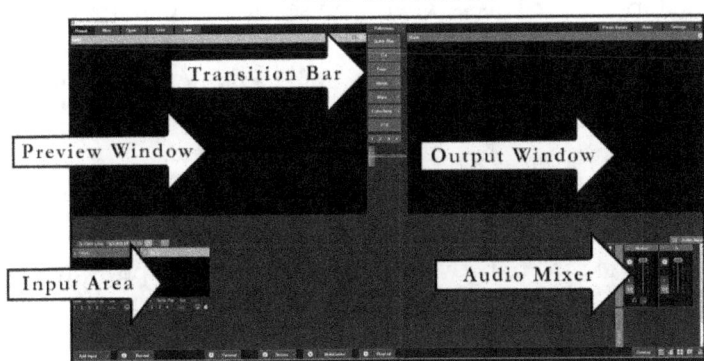

The Interface

vMix is packed with features and options. Fortunately, you do not need to master all of them to get started. It is best to first familiarize yourself with the overall layout and functionality of the interface.

The Preview and Output Windows

When you open vMix for the first time, you will see a blank canvas. This will come to life as you begin to add pieces to your production. Taking up much of the screen, you will see two windows. The on the right is the "output" window, and the one on the left is the "preview" window. The output window shows what is currently being sent to your live stream or recording. This is often referred to as the program monitor. Whatever is in the output window is "live" meaning its being recorded or live streamed depending on the enabled features. The preview window lets you view an input source before sending it to the output with a transition button. Using the preview window, you can be sure that you have the right source, and everything looks good before switching it to the output.

The Input Area

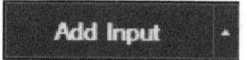

On the bottom left of vMix, is one of the most important parts of the interface, the input area. This is where you set up everything you want to add to your production, including cameras, videos, images, and websites. Anything that you eventually want to end up on the output screen needs to start here.

The Transition Bar

In the center of the screen, between the preview and output windows, is the transition bar. This is where you can switch from what is in the preview window to the output. There are multiple options for this, including an instant cut. Next to each of the other additional options is a dropdown menu where you can choose from several additional preset transitions. There is also a manual fader bar at the bottom that allows you to manually control the transition speed.

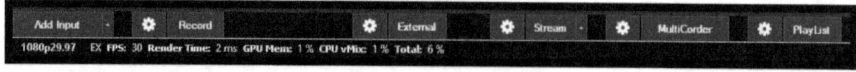

Other Features

Other features of vMix will be covered in future chapters but you should take some time to explore the interface. Check out the Settings menu in the upper right to adjust some of the more advanced settings. Also, look at the control button option across the bottom bar next to

Add Input. Record enables recording of the output. Click on the gear icon to choose the filetype, location, and other recording settings.

The **External** output feature allows you to send audio and video out of vMix in several ways. Click on the gear to see how you can select what sources are assigned to what outputs. Click the gear next to the stream to set up your live stream to multiple destinations.

On the far right are several menus, including one for overlay settings, statistics, shortcuts, and a hamburger menu that will reveal several other submenus.

There is no need to master all of these menus and settings yet. This chapter is to help you understand how the interface is laid out and how to find things as you need them later. Very few operators will ever use every possible setting or feature of vMix, but it is good to know where they are.

Adding Inputs & Input Settings

Inputs are the foundation of your vMix production. An input is any element that you want to add as part of your video production. They can include cameras, videos, images, PowerPoint presentations, audio and other sources. vMix makes it easy to add inputs and then adjust a wide range of settings to get your input ready for your production.

Start by adding a camera. Most video productions use cameras as the central element connecting viewers to the event and activity. Most cameras can be connected to your computer via capture devices, either with internal PCIe capture cards or external SDI or HDMI capture cards via USB. Cameras can also be connected over an ethernet network using NDI.

To get started, make sure your camera is connected and click on the "Add Input" menu at the bottom left of the screen. When the dialog box opens, you will see down the left side every type of input you can add. You will learn in more detail how all these different sources function. For now, click **Camera**.

Adding Inputs

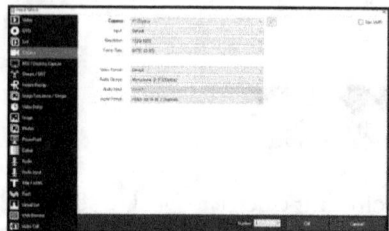
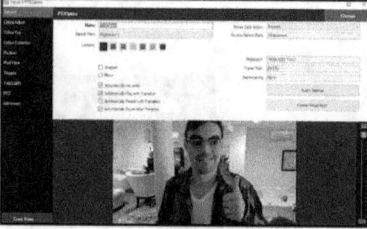

Selecting an Input   **Settings for Inputs**

From the dropdown menu, click on your camera or capture device. If you are using a Magewell, Blackmagic, or AJA capture device, vMix should be able to automatically detect your settings from the capture device and camera. You can double check those settings in the dialog box. You will want to be sure the camera's frame rate matches the master frame rate of your production. You can double check that in the corner of your interface screen. If you would like to use the audio from this source, click on the **Audio Enabled** checkbox and choose the source from the dropdown menu. When you are done, click **OK**.

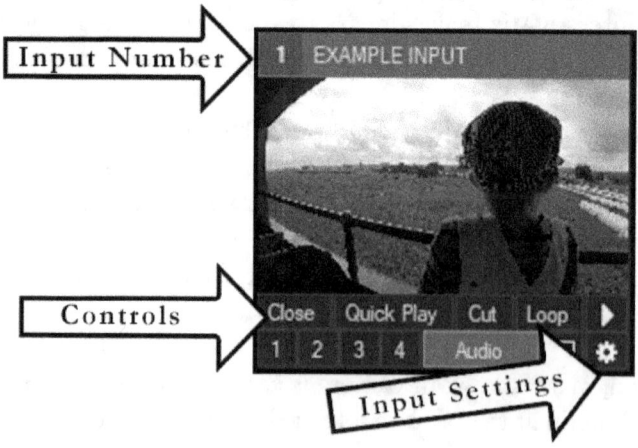

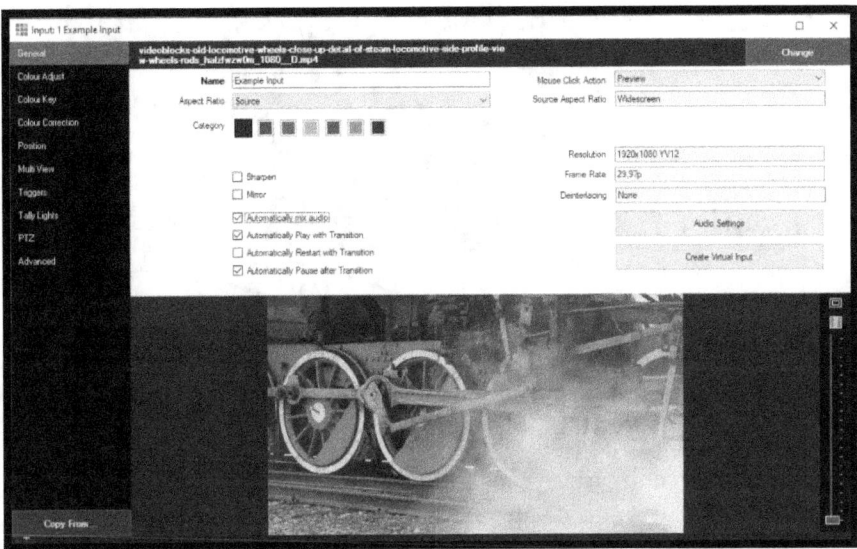

Once you have added a source, you will have access to many other settings. Just locate the input and click on the gear icon. This will open the settings menu for that source. You will likely want to change the name of the source here so it will be easy to reference as you add more sources. You can also add the source to one of the color categories. Color categories can be used to organize your inputs into tabs. As your production grows, you should consider naming every input and organizing inputs into these tabs for quick reference. You can create names for these tabs by right clicking the tabs and entering titles.

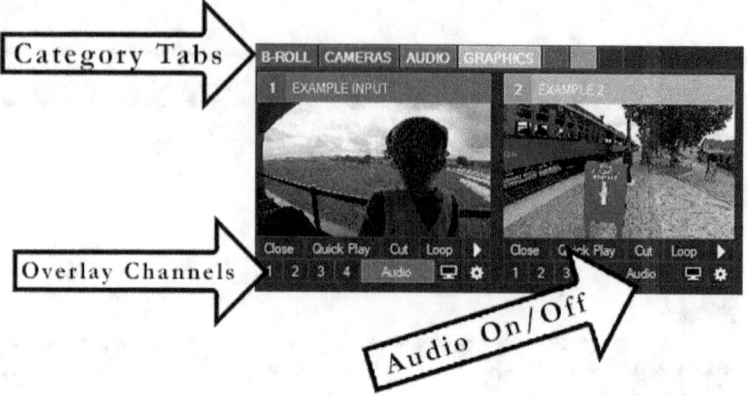

Under general settings you can determine what will happen when you click the source. You can have it move to the preview window, cut directly to the output, or choose from transition and overlay options.

☑ Automatically mix audio

☑ Automatically Play with Transition

☐ Automatically Restart with Transition

☑ Automatically Pause after Transition

You can also choose to to check the **Automatically mix audio** setting. If your individual cameras are connected to their own microphones and you want that camera's microphone to become active when you switch to the camera, check this box. If you have a master audio source from a sound board of independent mics, you will want to be sure this box is not checked.

Down the left side of this dialog box you can explore the various other setting options you will have for this input. You can adjust the color, apply color correction, set position, set up chroma and luma keys. You can also set multi-view options, set up action triggers, control tally lights, and even remotely operate PTZ cameras.

Once you have added a camera, you can go ahead and add additional sources following the same steps.

Mixing Inputs Together with MultiView

vMix's Input MultiView makes it easy to design custom layouts that combine multiple input sources into one scene. For instance, you could have a camera input as the main image and then add a backdrop, an inset picture-in-picture video, a lower thirds title, etc. The best part is that, when you create this MultiView input, all these sources are still usable as independent sources.

Input MultiView is not the same as overlay channels. You can setup a Multiview input combining several inputs and still have access to all the overlay channels. Also, keep in mind that Input Multiview is not the same as the Fullscreen MultiView output, which allows you to see and organize views of your entire production. For example, Input MultiView would be used to create an input with multiple layers and a Multiview output would be used to monitor your production on a second monitor.

Input Multi View vs Fullscreen Multiview

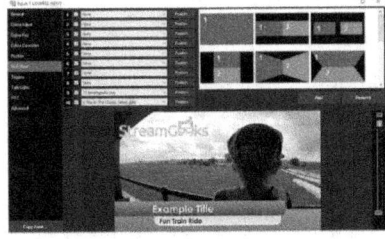 

Input Multi View   Fullscreen Multiview

Creating a MultiView Input

Up to ten MultiView layers can be set up on any input. A great way to get started is to create a brand-new blank input for testing. Just click **Add Input** and select the **Colour** option. Next, click the gear icon in the newly added input. Under the **General** options, give it a name so you can easily find it later.

Input Multi View

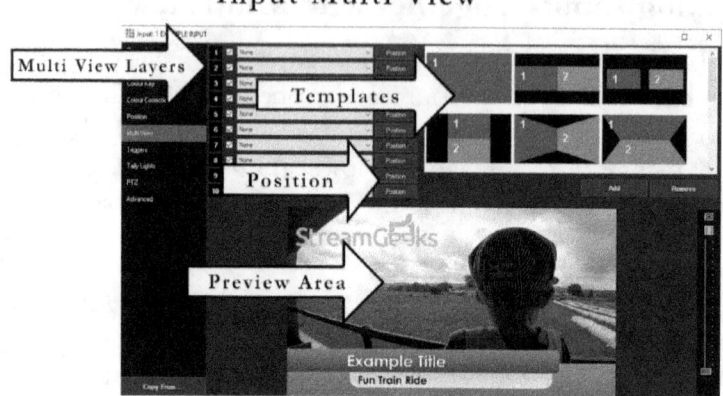

Setting it Up

Now select **MultiView** from the options on the left side to begin setting it up. On the right, you will see various templates for your MultiView like side by side, picture in picture, and a three by three grid. These templates are just a starting point to save you time and effort. You can change the layout as you go, and even save layouts for the future using the **Add** button.

On the left are the ten positions where you can add inputs to your Multiview. The order represents the position of the layers back to front. Input Ten is the top layer, whereas input One is the furthest layer in the back.

Once you have selected all your inputs, there are a couple of options for moving them around. You may be satisfied with the layout of the template you choose. But, if not, you can click and drag to move the elements around in the window at the bottom. Holding the shift key while clicking and dragging will resize the input keeping the aspect ratio in scale. To gain more control, you can click the **Position** button next to each layer and use the sliders to adjust the layers position.

Saving Templates

Once you have a layout you like and think you may want to use it again for other MultiView setups, you can save it as a template. Just go back to the templates in the upper right and click **Add**. When you load a new template it basically imports the settings you have created for each layer. So, as you add new inputs to each layer, they will automatically include the Zoom, Pan, Rotation and Crop properties saved inside of that MultiView preset.

Once you know how to create a MultiView from scratch, it is just as easy to create one from an existing input. This is great if you wanted to add something like a logo or title to an image or video input.

Once you have a MultiView created, you can even use them within other MultiViews. When creating a new MultiView and selecting the inputs, you can choose existing MultiViews from the drop-down menu.

There is really no limit to the creative ways you can use your inputs to make scenes and templates with multiple inputs on one screen. Even better, you still have access to your overlay channels for even more options.

Pro Tip: Use your overlay channels when you want to quickly show or hide inputs on screen. Use MultiView when you want to build a more complex scene that will be a base layer in your production that you can switch to.

Working with Titles

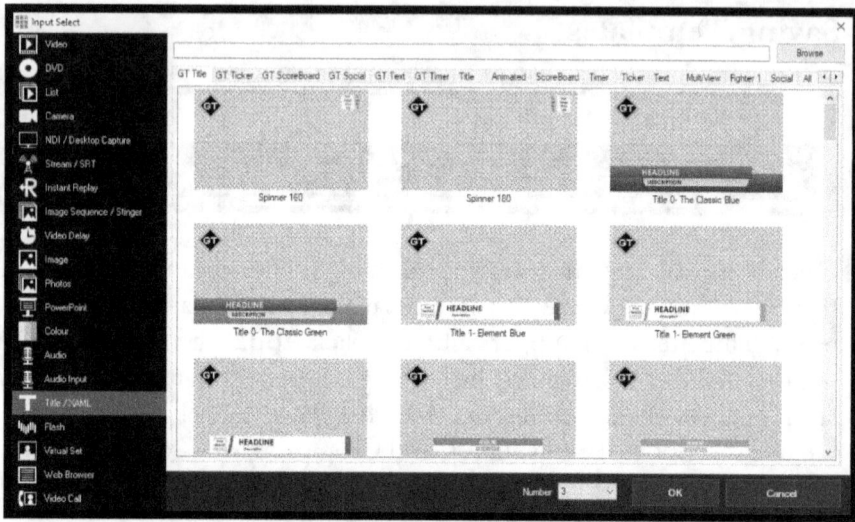

In an upcoming chapter, you will learn how to use the vMix GT Title Editor to create custom animated titles. But before you dig into creating custom titles, you should know it is incredibly easy to add lower thirds and graphics into your vMix production by using the included graphics in the **Title/XAML** input tab.

To get started simply click **Add Input** and select **Title/XAML**. Here you will be presented with a variety of title templates that you can choose from and customize. If the title has a little GT diamond in the top left corner that means the title has been built with the GT Title Editor and includes an animation. Select a title from the menu and click **OK**.

Pro Tip: You can open these titles with the GT Title Editor and customize if you wish. You can find them located in the vMix folder on your computer inside of the Titles folder.

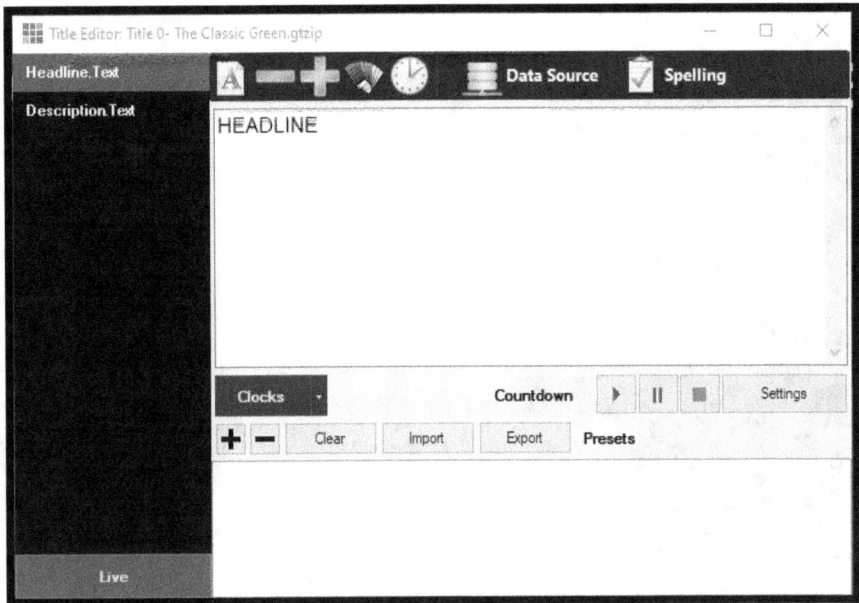

This will create an input inside of your vMix production for your selected title and open the **Title Editor**. You can reopen this Title Editor at any time by right clicking your title input. At the top of the Title Editor you have options for editing the text. You can alter the font, text size and color. The **Data Source** button allows you to map custom data from sources around the world directly into your title. Data sources will be covered in an upcoming chapter.

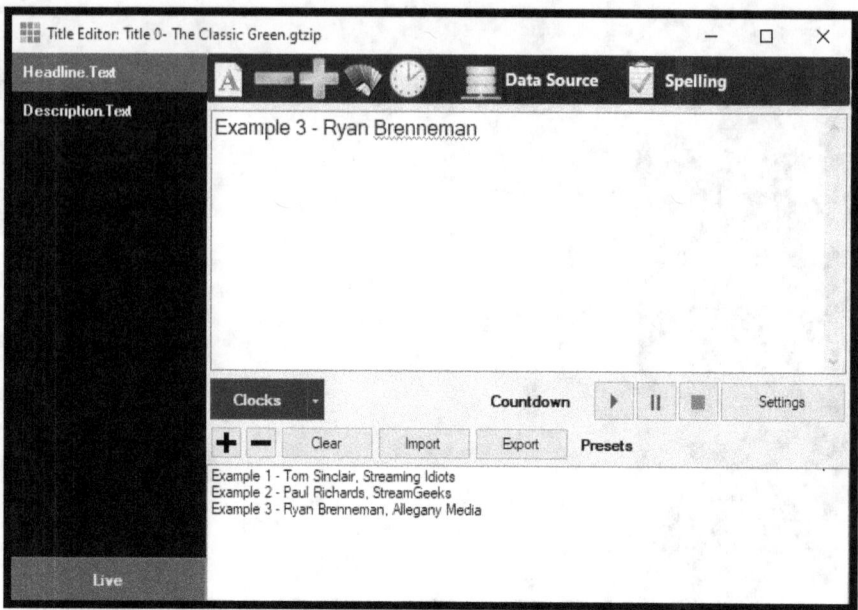

Many times, during a live production, titles have to be updated. vMix provides a list feature which can be used to build lists of information to quickly populate the titles when clicked on. You can use this feature by entering information into the title fields and clicking the **+** button. Each time you click the **+** button a new entry to the list will be added. You can click the **−** button to remove entries. In this way, you can quickly build a list and click the entry you would like to display inside of your title without having to enter the information in manually during your live broadcast.

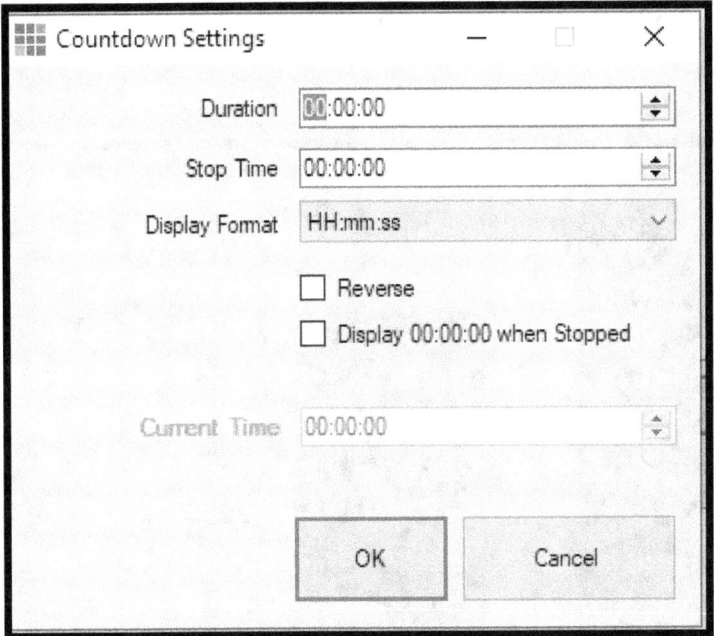

Some titles include countdown timers which can be set up using the **Countdown Settings** area. You can access Countdown Settings by clicking the **Settings** button next to the Countdown playback buttons. Many vMix producers connect Countdown timers with shortcut buttons to quickly reset countdowns during sports games for example. You will learn more about triggers and shortcuts for countdown timers in an upcoming chapter.

Most countdowns start with a specific duration time. For example, you can setup a duration of 30 minutes, click the play button, and your timer will countdown from 30 minutes. You also have a number of formatting options to customize your title. You can choose to show different combinations of hours, minutes, and seconds in your titles. You can choose to reverse your countdown and you can choose to display the current time using the **Clocks** button.

Pro Tip: Try setting up a shortcut to start and stop a countdown in your title.

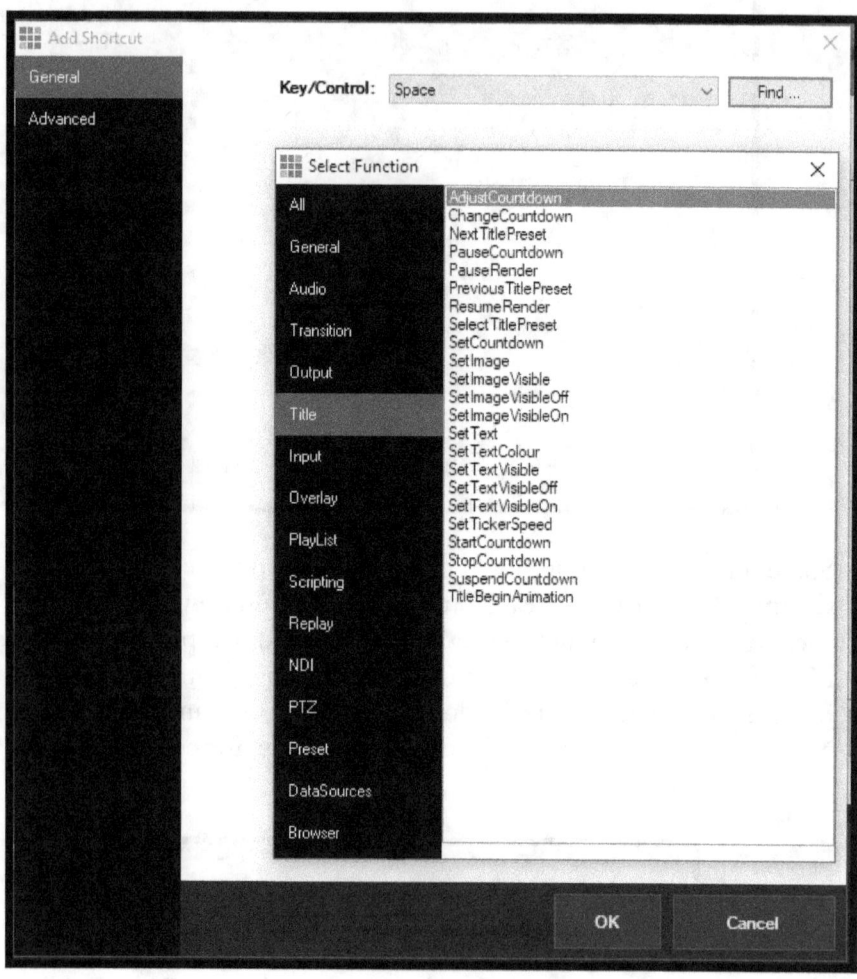

3 AUDIO

An essential part of any great video production is quality audio. Fortunately, vMix includes flexible and powerful audio tools and controls. To see how it all works, start by adding a dedicated audio source. Audio can also be included on all common video inputs. Still, you may also want to add stand-alone audio sources like microphones, audio interfaces, and mixers.

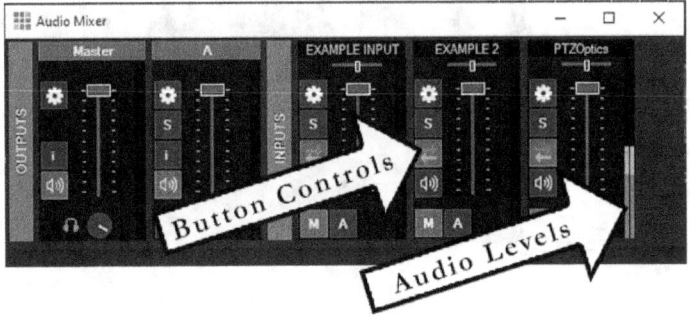

Add the Input

Click on **Add Input** and look toward the bottom of the left-hand column for **Audio Input**. Click on that, and you can select the live audio source you wish to use. From the dropdown menu, choose the source. Before you click **OK** and leave this screen, look for the **Mute in Headphones** checkbox. This is useful for any source that you do not want to monitor in your headphones. For instance, when using a live microphone to narrate a video production, some operators would rather not hear their own voice in the headphone mix.

☐ Mute in Headphones

Changing Settings

The next thing you may want to do is click on the gear icon on this new audio input source and go to the audio settings menu. The input settings are different from the **Audio Settings** but inside the input settings you can click **Audio Settings** to access the unique **Audio Settings**. You can also access audio settings by clicking the input settings cog inside the audio mixer.

Input Audio Settings

Indicator and Levels

Now, as you look at your input in the input source window, you should notice a green **Audio** button at the bottom right. That button will be green whenever the audio is on. If the audio source is sending a signal, you will also see that the levels next to the input are showing up as green bars and inside of the audio mixer. You can click on the **Audio** button to mute or unmute the audio. When you do this the button will turn gray, and the level meters will turn gray.

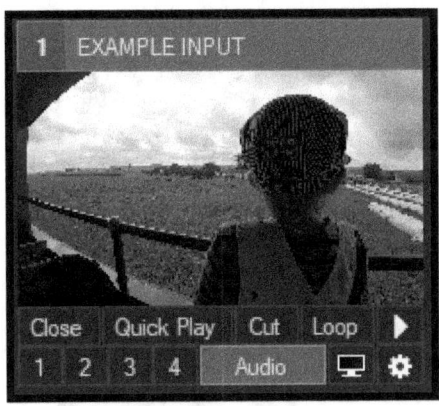

Inside **Audio Settings** you will have the option to adjust gain and delay for your source. You can also adjust the headphone volume or click the green headphones to turn headphone audio off for this input. There is a checkbox you can click remove the input from the audio mixer. This is helpful to clean up your audio mixer and only keep the input you need. The **Audio Settings** area also gives you controls over Gain, EQ, Compressor, Noise Gate, a Channel Mixer and a Channel Matrix. The VST 3 Plugins options will be reviewed in the next chapter.

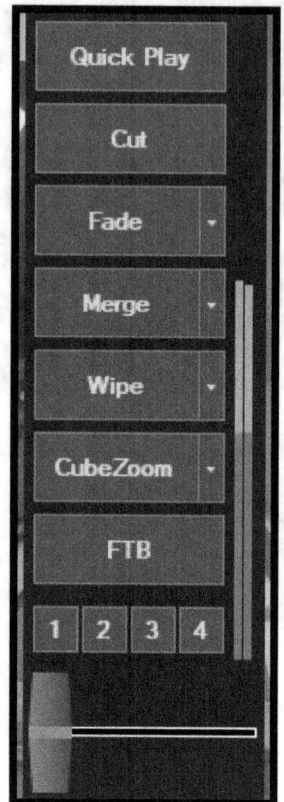

Master Levels

By default, all the activated audio inputs go to the master audio in vMix. If you have an active audio source, you will also see levels for the master audio mix shown next to the transition bar.

The Audio Mixer

By default, the Audio Mixer is located on the right-hand side of vMix. If you do not see it on the bottom right portion of your interface, look for the gray Audio Mixer button. If you click it, it will turn green, and the mixer will appear below it. Any input that you add will show up here with its own volume slider. It will have the same name as it does in the input window. This entire audio mixer can be undocked and moved. This is great if you are working with multiple audio sources and need more space.

The audio mixer enables you to adjust the volume of each individual source. You can also see the levels that will match those from the source over in the inputs area. Also, when you click the green audio button to turn off the audio on the input, you will notice the green speaker also turns off in the mixer area for that source, and the levels turn blue there as well. It works both ways. If you click the green speaker, you can toggle the audio from that source on and off. That will be reflected both in the mixer and on the input. This audio mixer is also where you will find the master audio for which you can also control the volume.

Other Options

Some other options found in the audio mixer include the settings cog, which gives you access to the advanced audio settings for that input. These include audio effects and control of the delay, which can be useful for syncing your audio with your video. Audio sync issues are common and vMix makes it very easy to apply audio delay to individual inputs in order to sync your audio and video.

If you have audio sync issues, it is most likely an issue that audio delay can fix. Depending on your computer and audio-visual connections latency between video and audio inputs can vary. Usually video sources can take just a little more time to process than audio due to higher bandwidth requirements. To fix this issue, try adding a delay to your audio inputs in increments of 25 until the issue is resolved.

Pro Tip: Included in the online course you will find an Audio Video Sync Tool. This is a video that you can use to determine the exact amount of latency you should apply to your audio sources.

Audio Mixer Controls

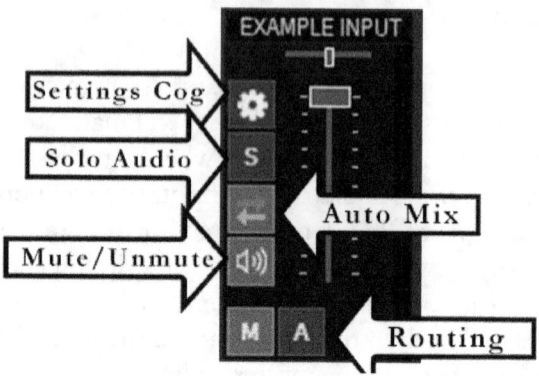

Below the gear is the **S** or Solo button. This allows you to listen to an individual source in your monitors or headphones, which is helpful when you need to hear one source separated from the other audio. Below the Solo button you will find the "Automatically Mix" button represented by two arrows. This button will automatically mute and unmute the audio of an input as it enters the output area. Next in the list you will see a quick mute and unmute button represented by a speaker. At the bottom, you will see an **M** button. That stands for Master. It means that this input is being routed to the Master audio for the production. If you have additional audio outputs set up you will be able to route audio to those outputs with these buttons. For example, you may want to set up a second audio output to send audio to a video conference software like Zoom.

Automatically Mix Audio

One of the most significant issues facing new vMix users is understanding how to control whether an audio source is active all the time or only when that video input source is active. For example, you may have some sources, like a host's microphone, that should be active all the way through a production. On the other hand, you may have a video with audio that should only become active when switched to the output window. vMix makes it very easy to choose how the audio for each source will function. It is also very easy to change the preference for each source. You can do this by going to the input settings for an input source and checking or unchecking the **Automatically Mix Audio** box. You can also toggle this on and off on the mixer by clicking the button with two arrows.

Audio Plug-ins

The ability to use third-party audio plug-ins give vMix users access to a massive collection of tools to help create professional-quality audio for their live productions. VST (Virtual Studio Technology) uses digital signal processing to recreate traditional audio hardware often found in professional audio studios. Now with vMix and VST, producers have access to the sound of high-end compressors, expanders, reverb units, delays, equalizers, effects, and more.

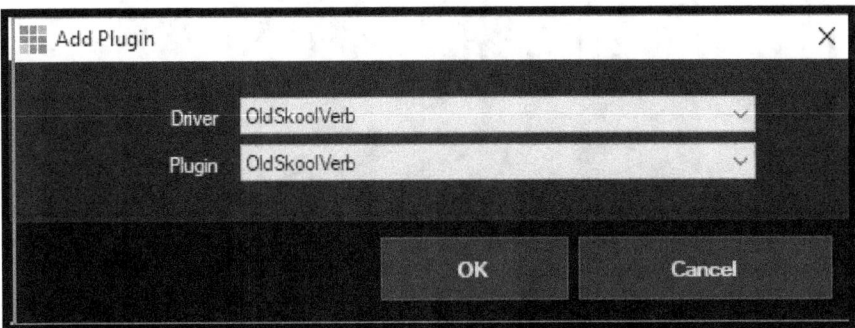

First Things First

Before you dive into audio plug-ins, there are a few things to double-check. First, be sure that you have vMix version 20.0.0.41 or higher. Also, keep in mind that vMix supports VST 3 64-bit plug-ins. You will find many VST and VST 2.0 plug-ins on the market, but for stability and reliability, look for VST 3 64-bit. They are available from many vendors, including waves.com. One of the most important things to remember is to carefully test these plug-ins before you go live with them.

Downloading and Installing

If you are looking for high quality and affordable VST 3 plugins, I highly suggest Waves.com. Two audio plugins I consider *must haves* are

NS1 (for noise suppression) and Renaissance Axx (A plugin to make voices sound more robust). Once you find a plug-in that you are ready to test out, go ahead and download it. Most downloads will come with an installer that will automatically place the files where they belong, usually: program files/common files/VST 3.

Now that the plug-in is installed, open up vMix, and you can see how to add it. You can add plug-ins to both individual inputs and to your master audio. There may be instances where you want to apply specific vocal effects to a microphone input such as reverb and a general effect to the master, such as a compressor.

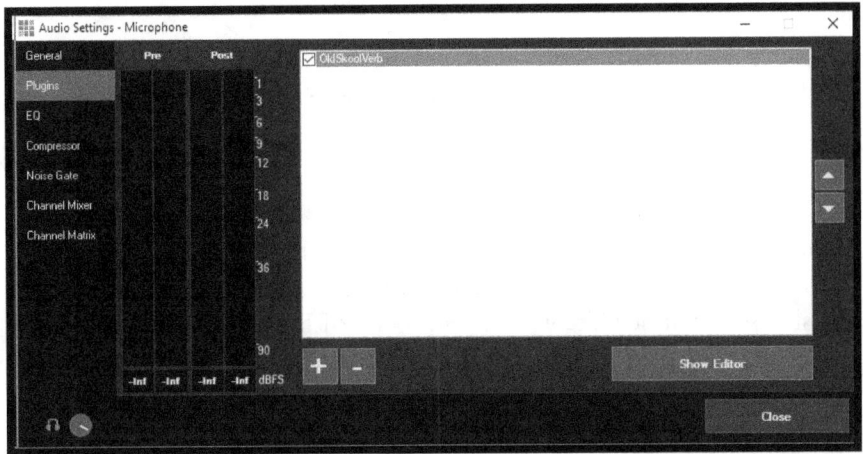

Go to the audio mixer and click on the gear icon on either the master or an individual input to open the **Audio Settings** "Plugins" tab. Now click on the + button to add a plug-in. Select a driver from the dropdown menu and then choose from any plug-in associated with that driver from the next drop down and hit **OK**. That will bring up the editor screen for that specific plug-in.

Different types of plugins have different interfaces. Some show a graphical interpretation of the physical version of the hardware, and others have simpler interfaces. Usually, once you add the plug-in, if that input has active audio, you should be able to hear it and adjust in real-time in the editor. Once you have it set up the way you want it, you can save a preset within the plug-in to quickly load it later.

Pro Tip: Many plugins come with presets you can use for various effect settings. For example, for voices there may be a male and female preset. This is often a great place to start your project.

Once you have your plug-in set up, you can make them easily accessible by setting up shortcut keys. You can even control them from a midi-controller, X-keys, or vMix web controller.

Audio Plug-ins are a great way to improve the sound of your live production. However, if you are new to them, they can be a bit overwhelming. If you want some guidance about what plug-ins work best with vMix, check out some of the online forums on vMix.com.

vMix Settings

The **Settings** menu, accessible in the upper right-hand corner of the vMix, gives you access to advanced features and settings. The most basic users will not likely ever visit many of these settings, but it is helpful to have a basic overview of what is available and where to find it.

On opening the **Settings** menu, you will see the 14 setting categories down the left-hand side. At the very bottom, look for the **Import, Export**, and **Default** buttons. These are you moving or duplicating your setup to another computer. Using **Export** will create a file with all the configurable settings. **Import** will load that file into another instance of vMix if it is the same version. **Default** will return all the vMix settings to how they were when the software was first installed.

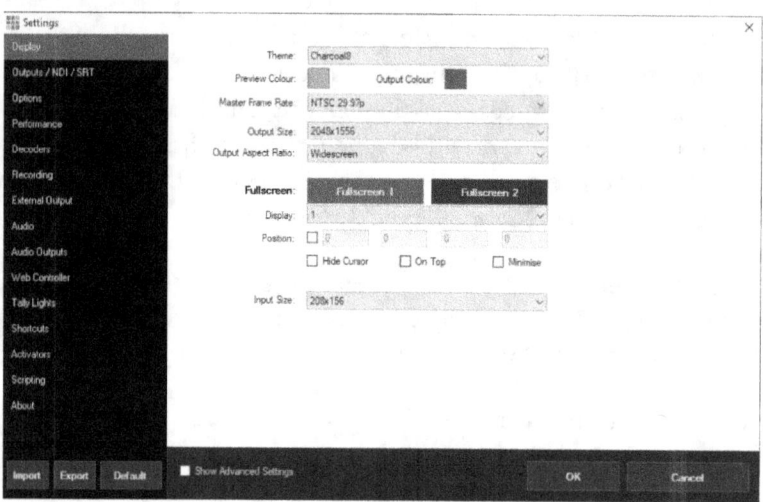

Here are the settings options available inside of vMix.

Display

These are the master settings for the vMix output. You can set the master output size, frame rate, and aspect ratio. You can also change the interface's theme colors and adjust the setting on the **Fullscreen** output(s). Here you can select the monitor vMix will use for your full screen output.

Outputs/NDI/SRT

This is where you can adjust the multiple vMix outputs, whether via your graphics card or over NDI. The number of outputs is defined by the edition of vMix you are running. You can choose sources for your full screen outputs here or via the "Fullscreen" button at the center area above the transitions bar.

You can also configure NDI (Network Device Interface) outputs which are IP video streams that are made available over your local area network (LAN). You can choose to turn on up to four unique NDI outputs in addition to enabling all camera sources, vMix calls, and audio inputs/outputs. In this way, you can make almost any source in vMix available to other computers on your network. By clicking the settings cog next to each NDI output you can configure the NDI outputs audio channel, resolution and video settings. You can even enable SRT outputs in this area with quality setting and latency options.

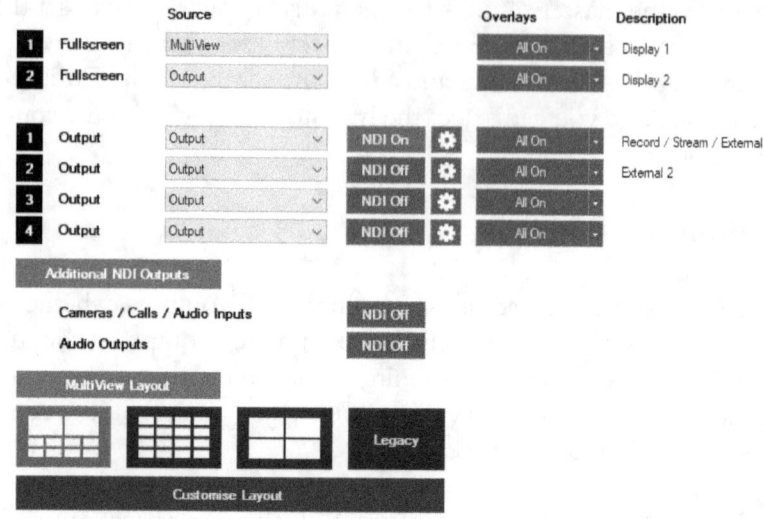

Options

This area includes many of the general preferences for vMix. These include the preferred language, the layout of the interface, and how the software behaves on startup and during a recording or streaming session. One new feature in **Options** will enable **Production Clocks**.

Production Clocks

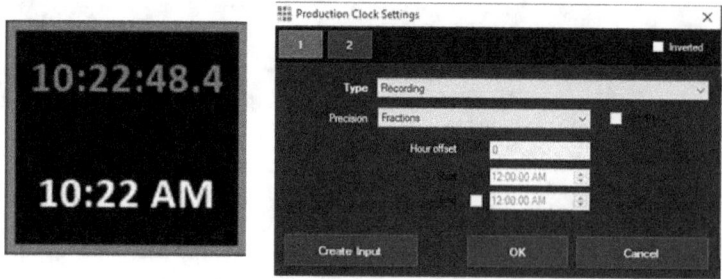

Once enabled a Production Clock will appear underneath the transition bar area. You can double click this box to bring up it's settings options.

Production Clocks allow you to quickly reference the time and a custom countdown time based on an event. You can also keep track of the time you have been recording or streaming.

Performance

The Performance settings tab is where you can select the graphics card you will be using with vMix. It is here you can enable **Low Latency Capture** and **High Input Performance Mode.** You can also access settings that will alert you when your CPU or GPU is pushing its limits. This is helpful so you can address any issues that may lead to poor performance or dropped frames.

Pro Tip #1: In addition to enabling your graphics card for use with vMix, check to see if your graphics card has been updated. For NVIDIA graphics cards for example, you need to open the NVIDIA Control Panel and add vMix as a selected program in the 3D Settings Area.

Pro Tip #2: If you are using a Windows laptop check out your battery settings. In "Power Options" you can enable "Performance Mode" which will greatly increase the performance of your laptop and its ability to run vMix smoothly.

Decoder

This section is for specifying what decoders will be used to playback certain video file types. In most cases, these can all be left at their default settings.

Recording

Here you can specify the default folder where recordings will be stored and the format for the filename. You can also set the size for the recording memory buffer. This may be useful if you have an older mechanical hard drive and are experiencing dropped frames during recording.

External Output

This is where you can access advanced settings for the available external outputs in vMix. Remember, most editions of vMix offer one external output, whereas 4K and Pro offer a second output. The External output can be used to enable a virtual webcam driver used to send video into additional programs such as Skype or Zoom.

Audio

This includes several preference settings, including a master setting for the Automatically Mix Audio setting. You can also turn off and on the audio meters for both the master and the individual inputs. There is an option to fade the audio entirely along with the video when Fade to Black in selected. Finally, there is a setting for a default audio delay across all inputs, which can be useful for dealing with syncing issues at a master level.

Audio Output

This controls the routing settings for the master output and headphones. You can also configure the seven auxiliary audio buses. Any of the outputs can be set to various output channels, including mixes for surround sound. Here you can also select the default audio bus for new inputs as they are added. Once you add an audio output it will show up as an audio routing option in the audio mixer.

Web Controller

This area contains everything you need to set up and use the vMix web controller. This is a web-based controller that you can use to switch vMix inputs, use shortcuts, edit titles, setup tally lights, and interact with the API. You can access this controller from any computer on your LAN using the web site address provided. You can optionally set up a username and password to restrict access to this controller. The Web Controller will be covered in more detail in an upcoming chapter.

Tally Lights

While there are other options for using tally lights, this section includes the settings for a tally light system based on Arduino hardware. Unless you are using such a system, you can ignore this settings tab.

Shortcuts

This is where you can set up, store, and edit shortcuts for nearly any function within vMix. Shortcuts can be assigned to the keyboard, MIDI device, and other control devices. Shortcuts will be covered in more detail in an upcoming chapter.

Activators

With activators, you can activate lights, buttons, and even motorized faders based on changes in vMix.

Scripting

For advanced users running 4K or Pro editions, this is where you can add and edit custom scripts.

vMix Full Screen and MultiView

There are lots of ways you can output video from vMix. You may be considering using vMix for live streaming or recording. Still, you may also want to send signals to projectors or external monitors. This could be for IMAG (image magnification) or simply to enhance production workflow by viewing the output, preview, or program views in an external monitor. In some studios, full screen outputs are used to power confidence monitors used by on screen talent.

Fullscreen

One way to handle external video is through the Fullscreen button located at the top center of the interface. There are many options for what is fed to this output. It can be connected to a monitor or projector through your computer's graphics card.

With vMix, it is recommended that you use a dedicated GPU (graphics processing unit). The software is optimized to take advantage of this additional processing power. It will take some of the workload off your CPU which can lead to better performance. These cards also tend to have multiple outputs. That allows you to take advantage of the numerous ways you can output video from vMix.

Before you get started with Fullscreen, connect the additional display to your graphics card and be sure Windows recognizes it. In the Windows Screen Resolution control panel, configure it to "Extend these displays."

With most production setups, you will have one output from your graphics card connected to a monitor to display your vMix interface for switching and control. With the additional screen connected, you can use the Fullscreen option and decide what will go to it. With vMix 4K or Pro, or the fully functional trial, you can add two Fullscreen outputs.

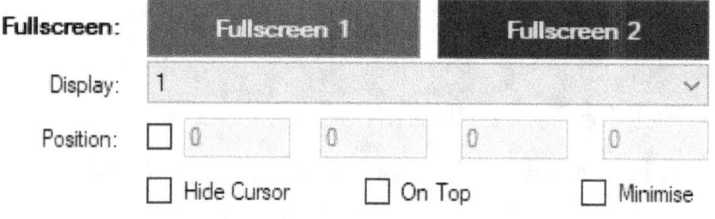

First, in the Settings section, click on **Display** and choose the display output you want to use for **Fullscreen**. Next, determine what you want to be sent to this display. By default, it is the final mixed output, but there are other options. Click on the down arrow on the Fullscreen button and choose from **Output, Preview, MultiView**, or any of the inputs you have set up in your input area.

MultiView

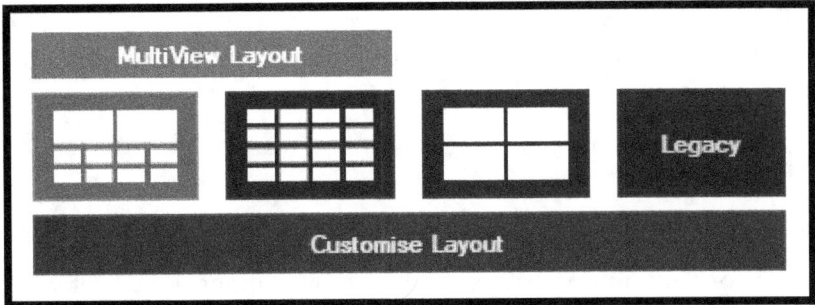

MultiView is an excellent option if you want an external monitor to display multiple input views. For instance, a typical setup might have the preview window, the output window, and eight commonly used inputs. Suppose another display shows up to 16 inputs at once. You could choose to have all your inputs displayed in the MultiView. To adjust what is shown in the MultiView, go to the settings menu, select **Outputs/NDI/SRT**. Under MultiView Layout, choose the general layout and then click Customize Layout to specify what should appear in each window of the MultiView.

Full Screen and MultiView are powerful features within vMix that give you a new level of flexibility and control. The more you dig into the features of vMix, the more important these viewing options will become.

Recording Video with vMix

vMix is a comprehensive solution for live production and live streaming. It also has powerful recording capabilities and can create numerous types of high-quality video files. Some users employ vMix mainly for the recording options. "Live to tape" productions are produced like a live show but recorded a broadcast in the future. Live event producers can record a high-quality file for backup, create other productions from recorded footage, or offer viewers an on-demand version after an event.

All editions of vMix offer numerous recording options. Still, with 4K and Pro, the functionality is expanded with options like MultiCorder, ISO recording of individual inputs, and instant replay.

Storage Space

One critical thing to keep in mind is that recording can use an enormous amount of disk space. It is an important consideration as you consider the type, quality, and length of the recording you will be doing. Be sure you have plenty of disk space, preferable on an SSD (Solid State Drive). Or investigate the external recording options that are possible with vMix. You can also setup additional computers to record outputs using NDI.

The main factors affecting recording files sizes are the format you choose and the bitrate you select. Some "lossless" video recording codecs such as AVI, create very large files. AVI and vMix AVI are the highest quality codecs and require the most storage space. MP4, WMV, and FFMPEG all allow you to adjust the bitrate to reduce the file size. By far the most popular codec is MP4. When you choose MP4, you can select the resolution, frame rate and bitrate. I have found that between 8-12 Mbps is great choice for quality and file size. Those who care very much about quality may increase the bitrates to 50-100 Mbps.

Pro Tip: vMix AVI is fault tolerant meaning that if you lose power you can still recover your video recordings.

Testing

It is also essential to test everything multiple times. This is especially crucial if you are working with new settings, recording multiple outputs, or recording and streaming at the same time.

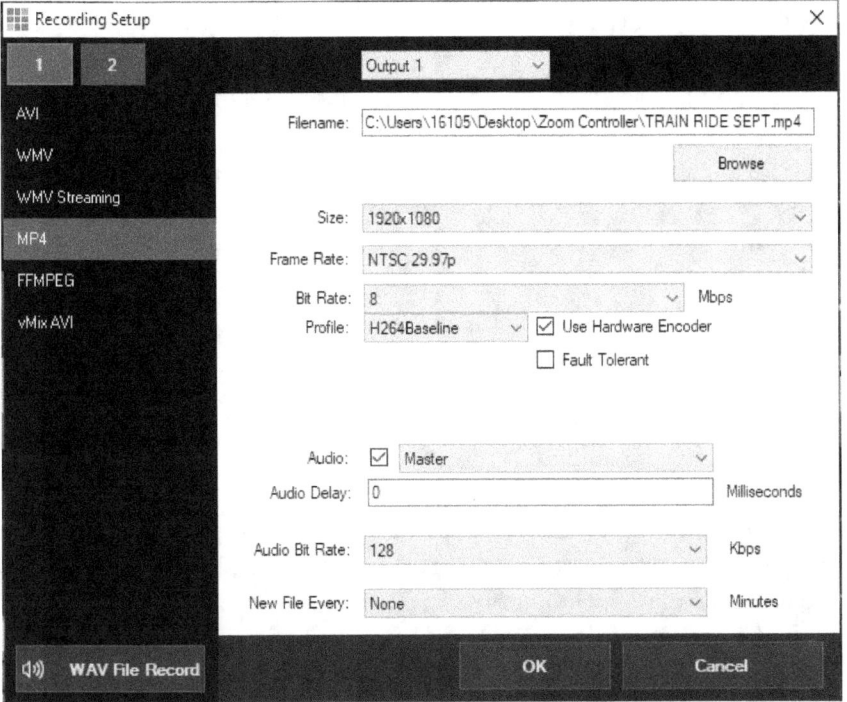

Recording Options

To see the many options available for recording, click on the gear icon next to the record button on the bottom left of the screen. Down the left-hand side, you will see the various file formats available to you. There are pros and cons to each file type. You will want to do some research to see what file type works best for your available resources and what you intend to do with the recording.

If you are not sure, or just want somewhere to start, MP4 is a widely popular option. It offers high-quality recording with reasonable file sizes. MP4 files can also be uploaded directly to video sites like YouTube or Facebook, with no conversion necessary. Even better, if you have a newer NVIDIA video card, the encoding can be done via the graphics card to free up some resources from your CPU. You can enable hardware encoding with the check box next to your recording profile.

Recording Settings

Once you have selected your filetype, you will see the multiple settings for the file. You can choose a location and name the file, choose the resolution, and set the bitrate. Below that, you can select the profile, which is especially critical if you are going to use your GPU for encoding. You will need to be sure the profile will work with your specific graphics card.

Further down, you can choose the audio source that will be recorded with the video. If you are unsure, leave this set to master. You can also set a delay for the audio if you need to do so to get it in sync and set the bitrate for the audio.

At the bottom, there are two important options. One allows you to start a new recording at a set interval of minutes. This is helpful if you are creating a long production and want more manageable files to work with. To the far left is the WAV File Record button. This will create a separate WAV audio file of your entire recoding. This is perfect is you just want an audio backup or intend to use standalone audio for a podcast or other audio medium.

Once you have all your recording options set, it is as easy as clicking on the **Record** button at the bottom. You can also connect this to any triggers, shortcuts, or manual buttons you want to use. The record button will turn red, and you will see a red REC indicator at the top of the screen. When you are done, just click the Record button again to end the recording.

vMix MultiCorder

The vMix MultiCorder is a tool you can use to record multiple files at the same time. The MultiCorder is only available in 4K and Pro editions of vMix. You can use it to record raw video and audio from any capture inputs available inside of vMix in addition to the main produced video recording which would include overlays and graphics.

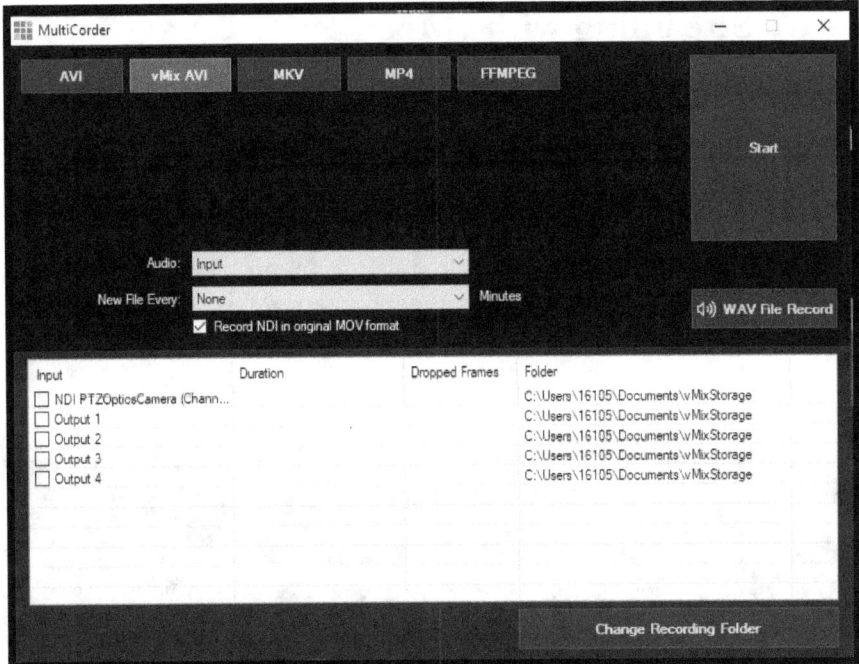

MultiCorder has the following minimum system requirements:

- Solid State Disk (SSD) for storing recordings.
- Intel Core i7 Quad Core processor or higher
- High end graphics card with at least 1GB of on board memory

You can use the vMix MultiCorder to record video in AVI, vMix AVI MKV, MP4 and FFMPEG formats. You can configure your MultiCorder by pressing the MultiCorder settings cog which is located at the bottom of the vMix interface. Here you can select the inputs you would like to record and configure the format you would like to use. Once you have configured the MultiCorder you can click it to start all of your isolated recordings.

Live Streaming with vMix

One of the most popular features of vMix is the ability to stream live productions over the internet. vMix makes live streaming simple and offers many powerful features and customization settings to help you get the most out of your live stream.

There is an ever-growing list of live streaming providers from free social media sites to paid professional content delivery networks. Each has its own feature set and its own technical requirements.

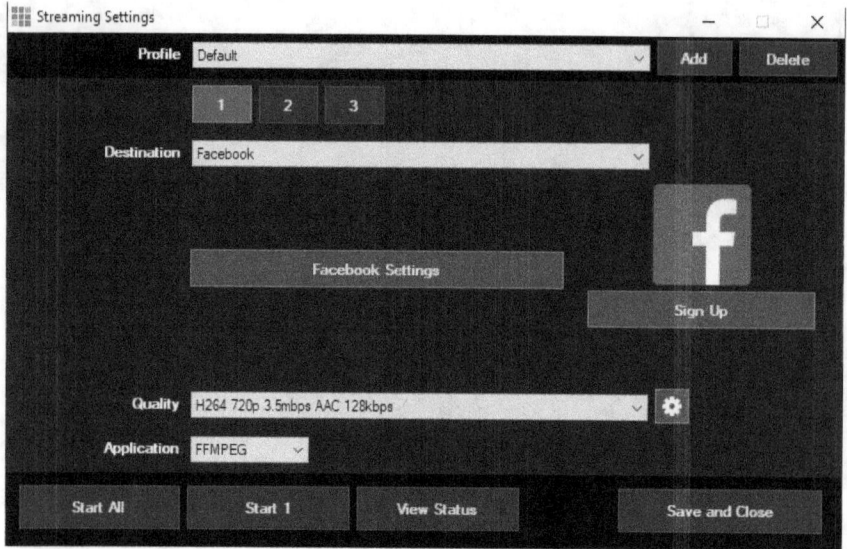

Getting Started

To get started, click on the gear icon next to the Stream button at the bottom of the interface. At the very top of the dialog box, you will see the Profile section. This will enable you to save and recall specific streaming settings. This is perfect is you find yourself using different providers and settings for various productions.

Right below that, you will see boxes for streams one, two, and three. Every edition of vMix enables you to stream up to three platforms simultaneously.

Choosing Your Provider

In the destination field, you will see an updated list of live streaming providers. You will likely find your provider on the list. However, since new providers are popping all the time, yours may not be on there. Do not worry, you can always set up your stream manually. To do that, choose Custom RTMP Server and enter the URL and stream name or key provided by your streaming provider.

If your provider is on the list, select it, and you can log into your streaming or social media account directly. You will notice that some of the dialog boxes are slightly different at this point. However, with most, you will just sign in and add some information. Some providers offer you multiple channels to stream to. Once you log in, these will be available from a dropdown menu.

Stream Settings

Once you are signed in and have entered any necessary information, you can choose your stream's quality. Depending on your edition of vMix, there are multiple presets ranging in quality from 360p to 4K. Most live-streaming platforms will give you suggested guidelines, and you are likely to find what you need on this list. However, if you need something different, you can choose any of the presets, click the gear icon, and change the settings.

When considering the quality of your stream, you will mostly be looking at resolution and bitrate. Your choice should be based on your provider recommendations, the quality of your production, and your internet connection's speed and reliability. Most streamers want their resolution to match the resolution of their production, especially when it comes to HD and 4K productions. They also want to use as a high of a bitrate as possible. However, this can lead to trouble if the internet connection is not fast enough.

Accounting for Network Resources

Your upload speed needs to be higher than your bitrate, or you will drop frames, and the stream could fail. It is a best practice to use a bitrate that is one half of your upload speed or less to leave room for network fluctuations. It is essential to know your internet speed and test it. When testing your internet speed, be sure to do so under similar conditions as when you will be streaming. For instance, if you are sharing a network with other users, you will want to test while using the network resources at the same level as your production time.

Bandwidth is measured in bits and the word "bandwidth" is used to describe the maximum data transfer rate of your internet connection. One megabit = 1,000 kilobits. Your internet speeds are measured in upload and download speeds. Megabits are used to measure the size of the bandwidth pipeline between your computer and the internet.

Resolution	Pixel Count	Frame Rate	Quality	Bandwidth
4K 30fps	3840x2160	30fps	High	30Mbps
4K 30fps	3840x2160	30fps	Medium	20Mbps
4K 30fps	3840x2160	30fps	Low	10Mbps
1080p60fps	1920x1080	60fps	High	12Mbps
1080p60fps	1920x1080	60fps	Medium	9Mbps
1080p60fps	1920x1080	60fps	Low	6Mbps
1080p30fps	1920x1080	30fps	High	6Mbps
1080p30fps	1920x1080	30fps	Medium	4.5Mbps
1080p30fps	1920x1080	30fps	Low	3Mbps
720p30fps	1280x720	30fps	High	3.5Mbps
720p30fps	1280x720	30fps	Medium	2.5Mbps
720p30fps	1280x720	30fps	Low	1.5Mbps

The chart above displays various bitrate/bandwidth choices you will have for your live streams. Using this chart and your available uploads speeds, you should be able to map out the number and quality of live streams your internet connection can support. A general rule of thumb says that you should only use half of your available upload speeds for live streaming (Download speeds don't help us with live streaming). Therefore, if you have 10 Mbps of available upload speed, you should only be live streaming with 5 Mbps. Leaving headroom in your upload speeds protects stream from fluctuations in the internet connection which can cause interference with your stream's consistency. Keep in mind that if live stream to multiple locations at the same time, you are adding to your overall upload bandwidth.

Think about your live stream resolution as the size of your canvas. The bitrate that you select is the amount of data that is used to fill that canvas. Therefore, you can have a high-quality 1080p stream with a bit rate of 6 Mbps, or you can have a low-quality 1080p stream with a bit rate of just 2 Mbps. Today, most people will expect at a minimum of 720p video and a bit rate of at least 1.5 Mbps. New reports from Akamai show that most people watching 1080p video find that 6Mbps looks like excellent quality.

So, you may have a choice between live streaming a single high-quality video stream, or multiple live streams of lesser quality. For example, if you have 10 Mbps of upload speed, you may create a 3 Mbps stream to YouTube and a 2Mbps stream to Facebook. If you are concerned about creating a single high-quality stream than you would only stream to YouTube using 5Mbps. Keep in mind that you can always record an high-quality recording to your local hard drive. Many production experts will record a "high bitrate" mp4 files ranging from 12-100 Mbps. The recordings saved to your local hard drive can be of higher quality than the live streamed recordings available on YouTube and Facebook.

Starting the Stream

When all the settings are the way you want them, you can click **Save and Close**. You can start your stream from that same window or just click the Stream button on the bottom of the screen. The button will turn orange as it is connecting and the turn to red. If there is a problem with the stream at any point, the indicator will turn to orange. When you are done, hit the button again, and your stream will end.

Pro Tip: If you are live streaming to multiple destinations you can start all of them at the same time by clicking the **Stream** button. If you want to start each stream at different times, open the Stream settings area and click the individual stream start buttons.

Video Overlay Channels

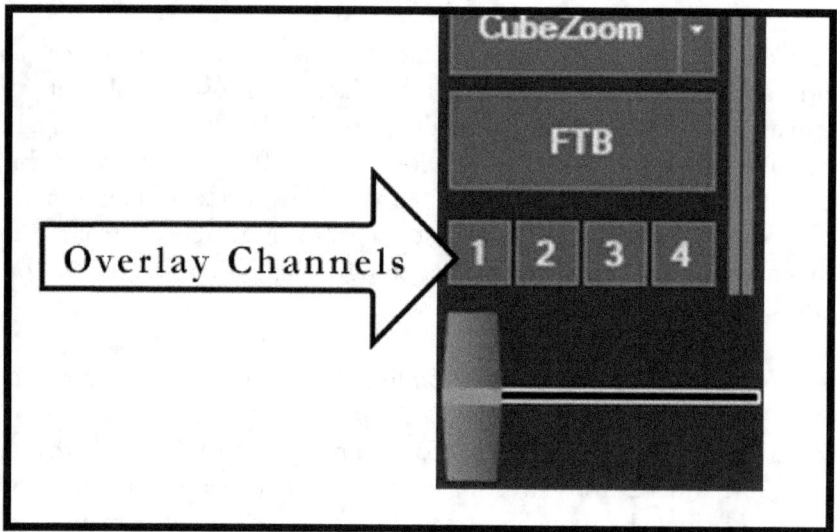

Overlays, available in vMix, are a great way to add depth and dimension to your live video productions. With the touch of a button or click of a mouse, you can overlay multiple graphics and other inputs over your main output layer. You can produce the same look as professional broadcast studios with lower thirds, picture in picture videos, and much more.

Getting Started with Overlays

To get started, be sure you have the graphic, video, or camera added as an input in the input area. You can add it to the output as an overlay by clicking on the selected overlay number at the bottom of the input box. Depending on the edition, there will either be one or four options. If you are using the Basic or Basic HD editions, you will have access to one overlay channel, HD and higher offer four. If you are using the free trial, you will have access to all four.

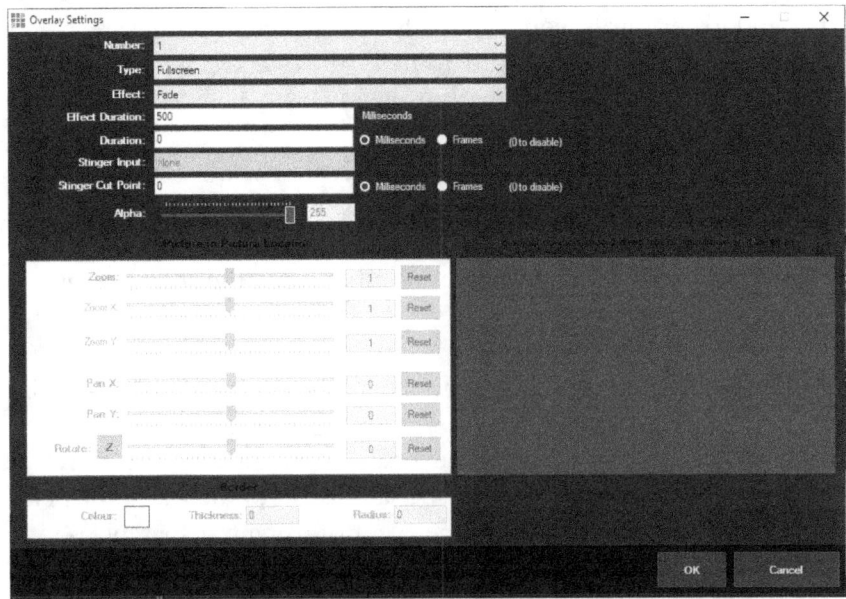

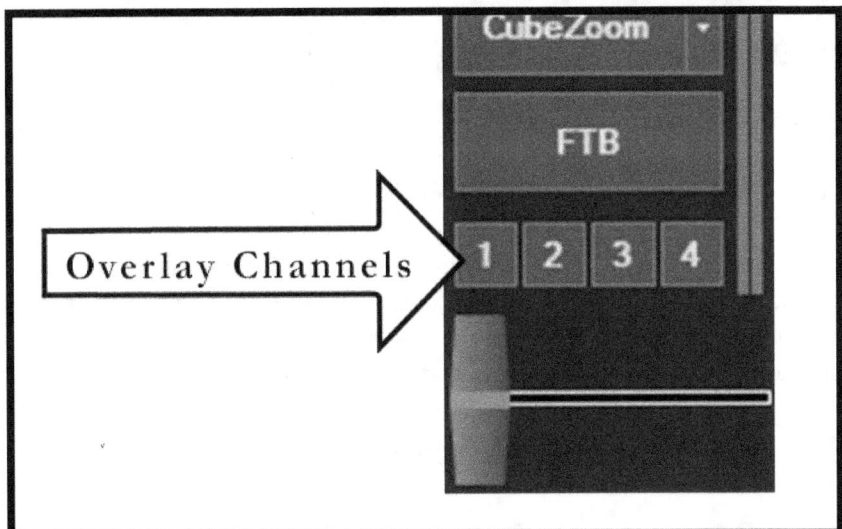

Overlay Behavior

To determine how each overlay channel will behave, click on the **Overlay** button at the bottom of the screen. This will open up the **Overlay Settings**. Choose the overlay channel you want to work with from the dropdown menu in the first field. Next, choose the type. For titles and lower thirds, you will usually want to choose full screen. You can also select picture in picture to make the overlay channel resize the selected input and scale it as part of the effect.

Next, from the **Effect** dropdown, you can choose a basic cut transition or choose from multiple transition effects. Below that, you can select the effect duration, which will determine how long the wipes, fades, or other effects will last. If you are using a basic cut, the software will ignore the duration setting.

Next is the duration of the overlay. This is how long the overlay will stay on the screen after it is activated. If you want to control this manually, just set this to 0. The next two settings are for the more advanced stinger transitions that add animation to the transition. The alpha setting is also available in the event of using transparencies in overlays.

For picture in picture options, the bottom half of the settings box lets you position your overlay element on the screen. You can adjust the size and position using your mouse, dragging to move the position, or clicking and dragging to adjust the size. You can also use the zoom, pan, and rotation sliders on the left side. At the very bottom, you can add a border and adjust its color, thickness, and radius.

Once you are done, hit **OK**. Now when you hit the number of that overlay on any of your input sources, that source will appear as you selected over whatever is in the main output window. It will match the transition settings, positioning, and duration you have set.

Mindful Video Production

As a vMix operator, it's important to think about your production from the perspective of the technical director. It can be helpful to think about your production as an orchestrated process of capturing the show's content with your audiovisual equipment. To do this, the technical director must make decisions about how and when to transition between the available video and audio sources. The goal of a good technical director is to produce a cohesive and captivating storyline. As you transition between one scene to the next, it is important to think about the viewer's experience as they follow along with your production. Choosing the appropriate moment with the appropriate transition type, is essential in capturing your audience's attention and making the story flow seamlessly.

Producers can use transition effects to complement their production capabilities. A good transition is transparent in the way it leads the audience through one scene to the next. In a perfect world, the transition happens as if the viewer naturally selected it themselves. To do this, your production should flow in a way that feels natural to viewers.

Be careful not to use too many fancy transitions that could take away from the main message of your content. Below are the four most popular transitions used in video production. In order of popularity, these transitions would be the cut, the fade, the fade to black and the stinger transition effects.

Type	Most Used
Cut	90% of the time
Fade	<5%
Other	<5%

You will notice that there are quite a few options in vMix when it comes to video transition effects. The most used video transition is a cut. The cut simply switches two video sources in a direct cut transition without any noticeable special effects. The cut should be used for transitioning between most of your prepared content. The cut is perfect for transitioning between two live camera angles in the same scene.

When you are cutting between multiple camera angles in one scene, it is important to think about the order in which you switch camera inputs. Jumping to too many different camera angles, too quickly could be disorienting for your viewers. A good technical director will visualize the camera angles that they have available and move through them in a natural order. You should try to arrange your cameras angles so that you can reveal additional details as they become more important to the story. If you have a pan, tilt, and zoom camera that can capture multiple angles during a single production, consider switching back and forth between close up angles and available wide-angle shots. Generally, you do not want to cut to a camera angle that is more than 45 degrees away from your current camera angle. In this way, you can switch between multiple cameras in an arch to finally reach a camera that may show a side or behind the scenes camera angle.

Interview Diagram #1

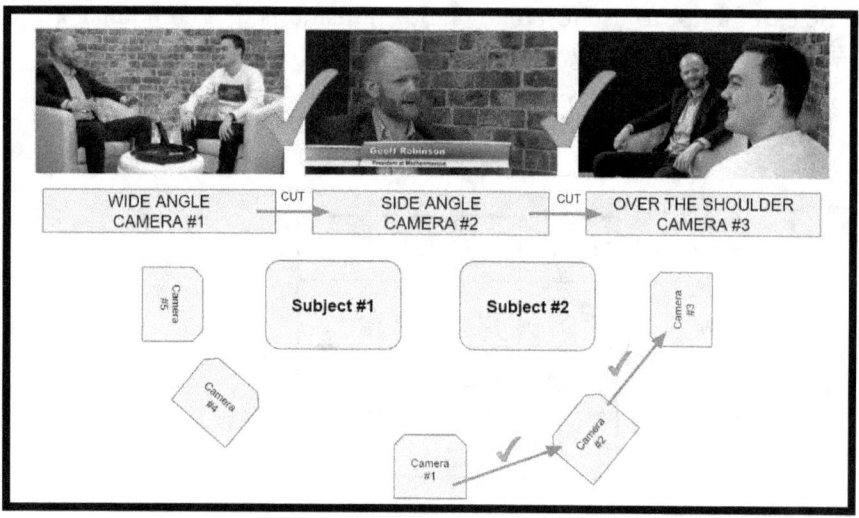

To study these production techniques, use the example of a two-person interview. When you are producing an interview like this, it is a good idea to start with a wide angle shot that displays both of your subjects in the same shot. This is your central shot that establishes placement for the viewers in the scene that they are watching. The cut transition can then be used to enhance the viewer's perspective of each person as they take turns talking. A transition like a fade or a stinger would look unnatural during a two-person conversation. It would make sense at the end of a conversation when you cut a commercial break. The cut makes the camera switching feel natural and unnoticeable because it happens in the blink of an eye. Each transition should be timed to flow with the conversation your subjects are having.

Interview Diagram #2

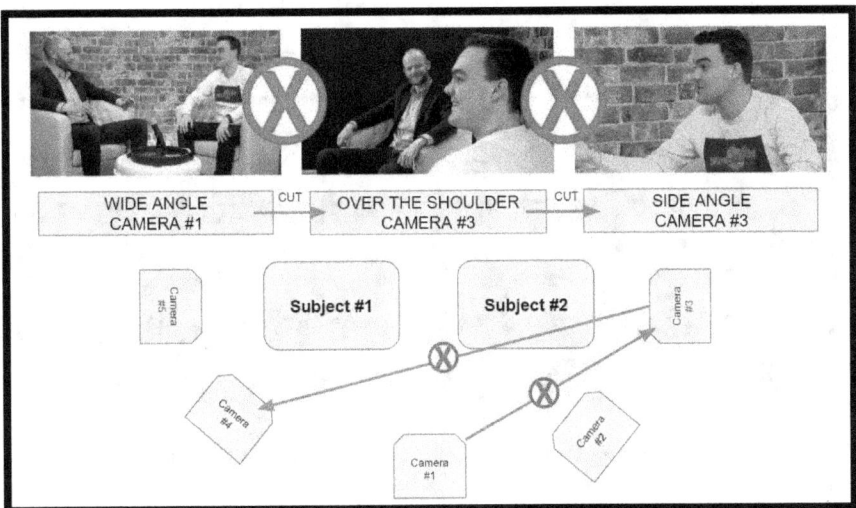

In interview diagram #1, the producer selected camera cuts that are sequential around the arc of camera options moving counterclockwise. Interview diagram #2 there is a sequence of transitions that do not follow sequential movements around the camera inputs. Diagram 2 breaks the 45-degree rule and risks disorienting viewers.

Using these types of "jumpy cuts" usually results in an unprofessional looking production. Remember that all rules are meant to be broken, and every camera setup is different. Use your instincts and create a production that makes sense for the story *you* want to tell. Just remember that the transition decision are an important part of video production.

The Crossfade

The crossfade transition is perhaps the second most popular transition in live video production. You will notice it being used less frequently than the cut. Crossfade transitions can produce beautiful artistic visual effects and should be used to create visually engaging content. Crossfades are used frequently during musical performances. Think about the national anthem performed just before a sports event. You will notice the crossfade is used when the cameras switch between a close-up of a singer and a wide panning shot of a crowd. You will notice great crossfade transitions that feature flags slowly fading into a crowd of singing sports fans.

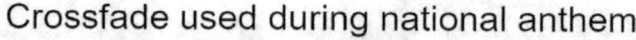

Crossfade used during national anthem

 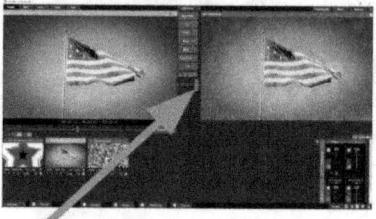

Crossfade t-bar position at 50%.   Crossfade t-bar position at 90%.

If you have an artistic shot prepared for your next video production project, the crossfade may be the most appropriate transition to use. If you have a pan, tilt, and zoom camera, try transitioning with the crossfade when the camera is in between a slow pan. You will learn more about automated PTZ camera controls inside of vMix in an upcoming chapter. Many professional broadcast studios use a physical T-bar to create custom cross-fade transitions between multiple video inputs. Use this transition sparingly and note that crossfades may look pixelated in low bitrate bandwidth streams.

Pro Tip: You can connect USB T-Bars to your Windows computer and program them to control the crossfade transition in vMix. Check out the XKeys XKE-124 TBar.

The FTB (Fade to Black)

The fade to black sometimes shortened to just "FTB", is perhaps the next most used video transition. You can fade to black and fade from black to notify your audience of the end of your production. This type of video transition clearly demonstrates the nonverbal communication power you have as a producer to close a scene. Try using the fade to black transition to close a unique scene or segment. If you can time your fade to black with the ending points of an audio track, or fade audio down it will help emphasize your production.

Using Stinger Transitions

Finally, the stinger transition is an animated transition that everybody has grown to love. The effect combines a transparent video animation that evolves into a full screen transition which is timed with a cut. When your stinger video animation starts playing on top of your current video, you can program vMix to cut to your preview input exactly when the video completely covers up the current scene.

The stinger effect has been made popular by sports broadcasters who use the effect to notify the audience of a particular scene change. In this way, sports broadcasters have trained their audience to expect a stinger transition when an instant replay or prepared video screen is coming up. Really fancy stinger transitions often include perfectly timed audio which is included with lots of "whooshes" and "pops" to give the animation a realistic look.

Pro Tip: Use the Multiview feature inside of vMix to add audio to your Stinger Transision. Click the "Automatically Play with Transition" and you can combine audio with your stinger transitions.

Making a Stinger Transition

Stinger transitions can be made with "**vMix Video Tools**" which is a Windows application installed with the main vMix application. This application is used to create an image sequence from a transparent video file. This method of creating stinger transitions is an efficient way to manage stinger transitions because the vMix Video Tools to break down a transparent video file into a sequence of PNG files which is smaller in size and more efficient.

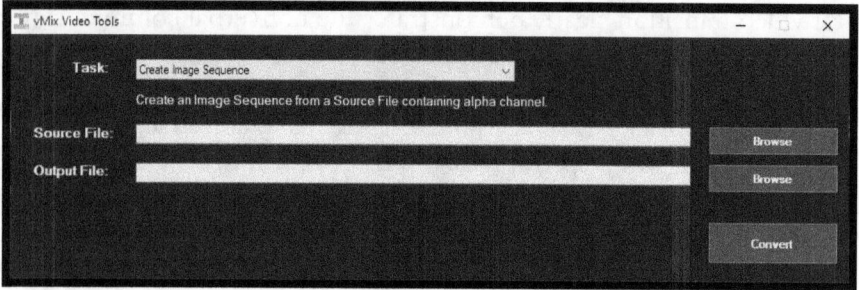

To create a Stinger transition, start by looking for a video with a transparent alpha channel background. MP4 files will not work. You should be looking for MOV or AVI files. Videoblocks.com is a great place to look for video content. Or you can create your own stinger transitions with a video animation software such as Adobe After Effects. When selecting a stinger transition video file, looks for an animation that eventually takes up the entire screen. This is the point where you can set up your stinger to transition to your preview input without the audience noticing.

Once you have a video selected you can select "**Create Image Sequence**" inside of **vMix Video Tools** and select your **Source File** and **Output File**. Once you load in your video, click **Convert**. This will create a folder on your computer with your Image Sequence that you can use to setup a Stinger inside of vMix.

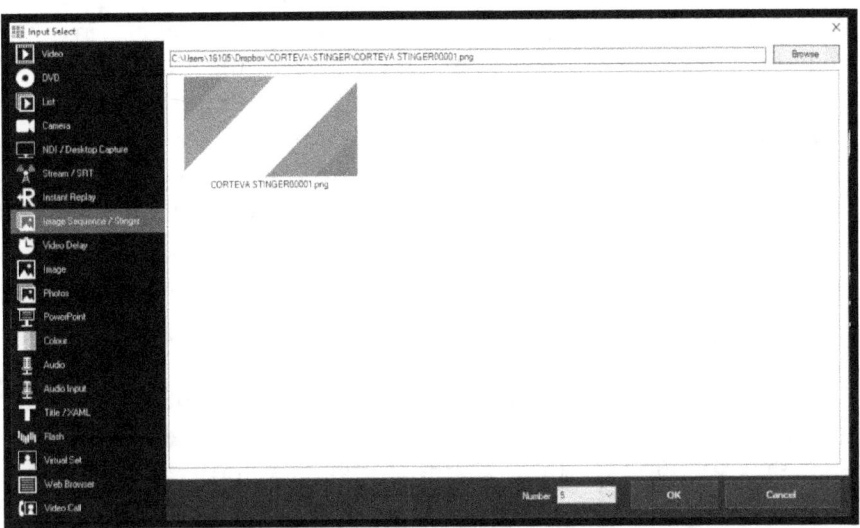

Inside vMix, you can **Add Input** and selecting the **Image Sequence/Stinger**. Once your image sequence has been loaded into vMix, you can click into the overlay section of vMix to set up the stinger details. vMix allows you to set up two unique Stingers in a single production. The most important detail inside the stinger options is the stinger cut point. You can test out your stinger a couple of times to tweak its performance to match the cut point where your animation covers the entire screen fully. In the overlay settings area, you will find multiple tools allowing you to set up the stinger duration, cut points and even its alpha channel properties.

Using Multiple Channels

You can use all four overlay channels at the same time. For instance, if you wanted to add a lower-thirds title, picture in picture video, a graphic, and even another video, you can use all the channels. For even more flexibility, you can use overlay channels with input MultiView sources. You can create a scene of up to ten layers and then assign that to an overlay channel.

Other Features

There are a couple more features of the overlay channels you may find helpful. They both work via a right mouse click. To see and overlay in the preview window, right-click the selected overlay channel on the input you wish to use. When use transition from the preview window to output, the overlay will transition as well. Finally, with a right mouse click, you can zoom a picture in picture overlay to full screen. Once you have clicked the overlay number to add the input as an overlay, simply right-click the same button to zoom it to full screen. Another left-click will turn the overlay off.

Using vMix Social

vMix Social is an easy way to bring social media content into your live production. With social becoming such a big part of events, these tools will save you time and make sharing social media posts and reactions an integrated part of your production.

Getting Started

To get started, go to **Add Input** and add a new title. Once you have selected the **Title** tab, you will see all the available title layouts. Click on the **Social** tab at the top. You can create your own custom social media title layouts, but you can also use one of the available templates. This will become the placeholder for the information that will come from vMix Social.

Once you select one of the social titles, you will see how they have been created specifically for social media. The title editor has ready-made fields for the type of information you will want to share from social media. These are placeholders for things like profile pictures, usernames, and social media messages. Some of the templates have a place for a title field. This is something you can enter in this editor, and it will remain the same no matter the data received from social media. Click on the X in the top right to close out of this window. If you want, you can click on that new title input and put it in the preview window to see what it will look like.

Connecting Social Media Accounts

Now it is time to connect to social media platforms to begin populating those fields. Open **vMix Social** by going to the hamburger menu in the lower right corner of the interface. It will automatically connect to vMix and show "Connected" in green at the top of the dialog box. Below that, you can see all the currently supported social media platforms. Keep in mind that these are subject to change. Social media sites are continually changing their access rules and processes, and some may become unavailable while others may grow in popularity and be added.

vMix Social Options

You can choose one social media platform or as many as you want. Clicking on each logo will give you access to the settings for that platform. You will see that each settings box is different since the various platforms have different login processes and search parameters.

Take Twitter, for example. To get connected, you will log in and then be provided a PIN by Twitter, which you will enter in the next field. Once you are logged in, you can choose your feed type. You may want to collect Tweets around a particular hashtag or search phrase, show a specific timeline, or even favorited tweets. Below that, you can enter your search term or username and make some other preference selections.

You can go ahead and repeat this process for any of the social accounts you wish to use. For Facebook, Twitch and YouTube, you can login to your accounts and select your specific live stream to receive comments from viewers. Once everything is set up and ready to go, you can add your social media title to your production using one of the overlay channels, keeping it in view all the time or only selecting it when you want it to be seen.

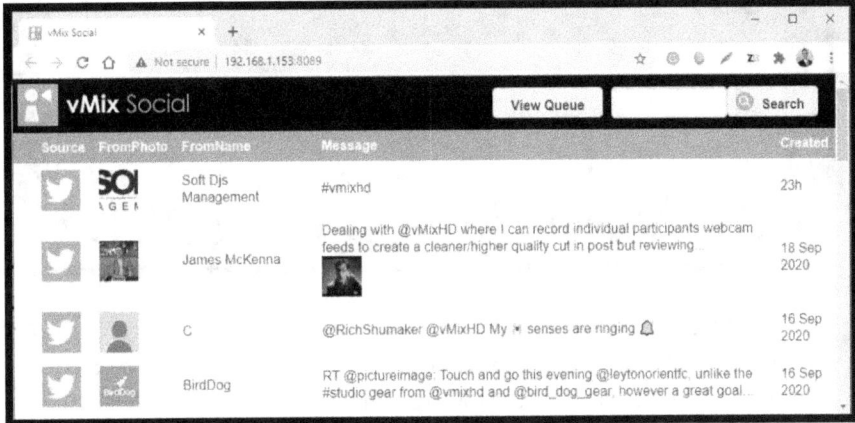

Comment Moderation

Of course, in most cases, you will want the opportunity to moderate social media posts and comments before they appear in your live broadcast. That is easy to do with the vMix Social web interface. Copy the URL from the vMix Social box into a web browser on any computer or device connected to the local network. The easy to use interface will allow you to select and approve content and send it directly to vMix, where it will appear on the social media title slide you created.

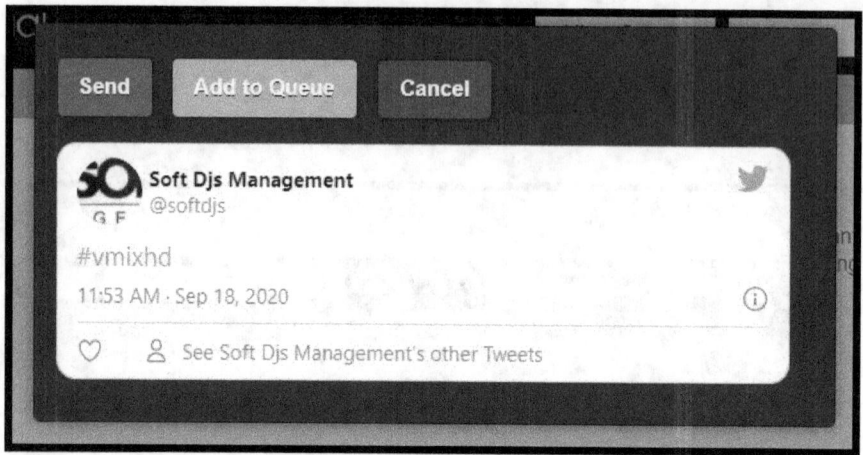

The specific title that will receive this information is the one that you select in the vMix Social application. You have the choice to **Send** comments directly into that title or build a list that you can manage by using the **Add to Queue** feature. Many times, you may receive great comments, but you are not ready to share them with your audience. Using the **Add to Queue** feature you can **View Queue** at any time to **Send** comments directly to your title from your curated list.

5 ADVANCED VMIX FEATURES

vMix Shortcuts

As your live productions become more complex, vMix makes it easy to access commands via your keyboard or MIDI device using shortcuts. Nearly anything you can do in vMix, you can do with just a keystroke or button press.

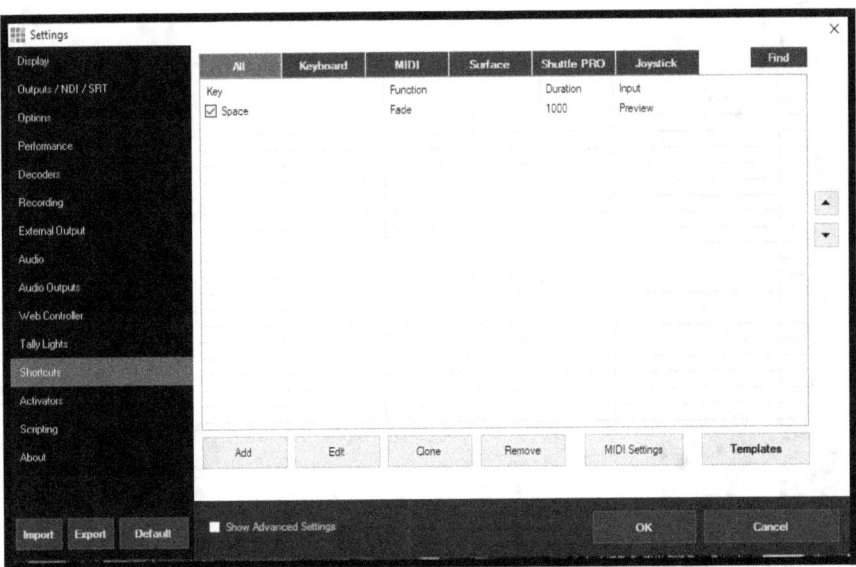

Getting Started

To get started, just click on the settings button in the upper right corner of your interface. Then select Shortcuts from the options in the left column. If you have yet to set up any shortcuts, you will see a blank screen. Across the top, you will see tabs for the multiple options for assigning shortcuts to different devices, including the keyboard, MIDI controller, control surface, ShuttlePROv2, or joystick. Start by learning how to set up a keyboard shortcut.

Setting Up a Keyboard Shortcut

Click the **Add** button at the bottom of the screen. That will open up the **Add Shortcut** dialog box. You can either type in the key in the Key/Control field or hit the **Find** button. If you hit the **Find** button, another box will open and wait for you to press a key. Once you do and it is correct, press **OK**.

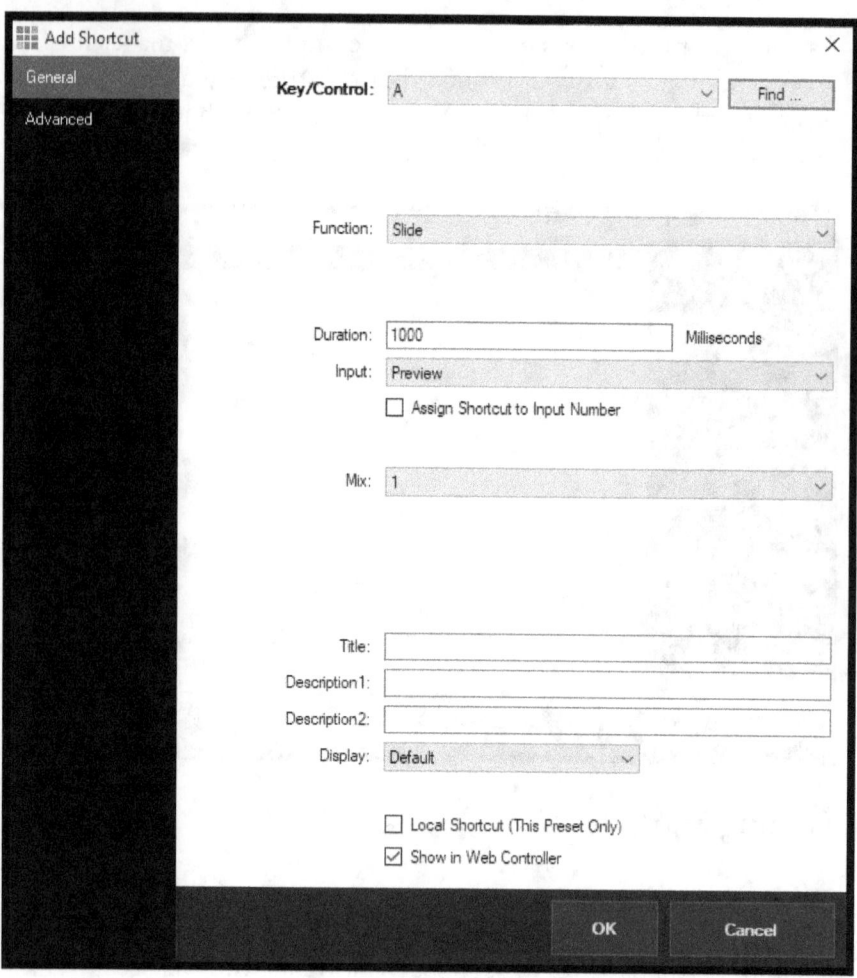

The next field down sets the function for that key. Click the down arrow to see all possible functions arranged by categories. Click to choose the process you wish to assign to that key. Next, you can select the duration for transition effects. Then, select the input you would like to apply the transition to. That can either be the preview window or any of the existing inputs. By ticking the box below, you can assign the shortcut to a specific input number.

You can then add a title and description to the shortcut and determine whether this will be a global shortcut available anywhere in vMix or local, assigned only to the current preset. You can also tick the box if you want this preset to be available in the vMix web controller. Then just click OK.

Once you have shortcuts set up, you can go to the shortcuts menu to add new shortcuts, edit, or even clone existing shortcuts. During the cloning process, you will find that you can assign multiple shortcuts to the same key. This is useful if you want to trigger multiple functions with one click.

Setting Up Shortcuts on External Devices

The process of adding shortcuts for a MIDI device, control surface, ShuttlePROv2, or a joystick is similar. Before starting, be sure that your device is connected and set up in vMix. From there, simply click **Add**. Next click **Find** and then press or activate the button you wish to use. The rest of the process is the same.

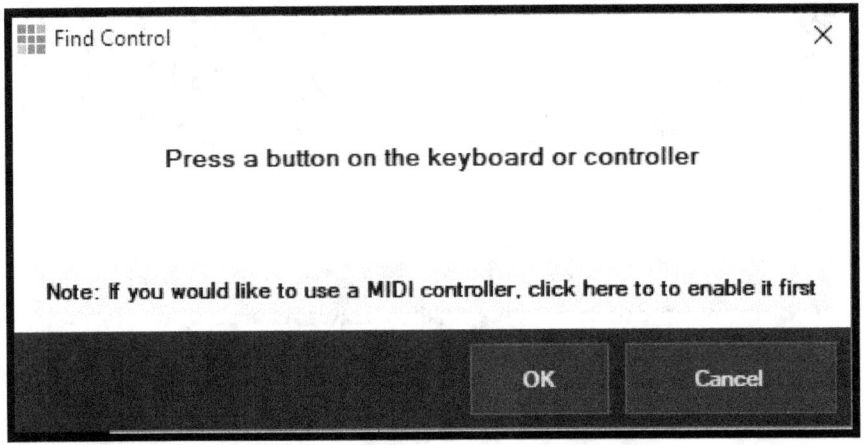

If you want to set up a lot of shortcuts at once, you may want to look at the templates found at the lower right corner of the shortcut's menu. There are multiple templates for keyboards, MIDI devices, and other controllers. They come preloaded with functions assigned to keys, but they are entirely customizable to fit your specific needs.

Using GT Title Designer

Titles can be an essential part of your live video production. They can identify guests, introduce segments, and share additional information related to your production. vMix GT Title Designer creates dynamic titles without overtaxing your computer's CPU. That is because the vMix GT Title Editor is GPU based. The designer is available in all editions of vMix. In 4K and Pro, you have the option to create custom animated titles and import Photoshop files directly in the PSD format.

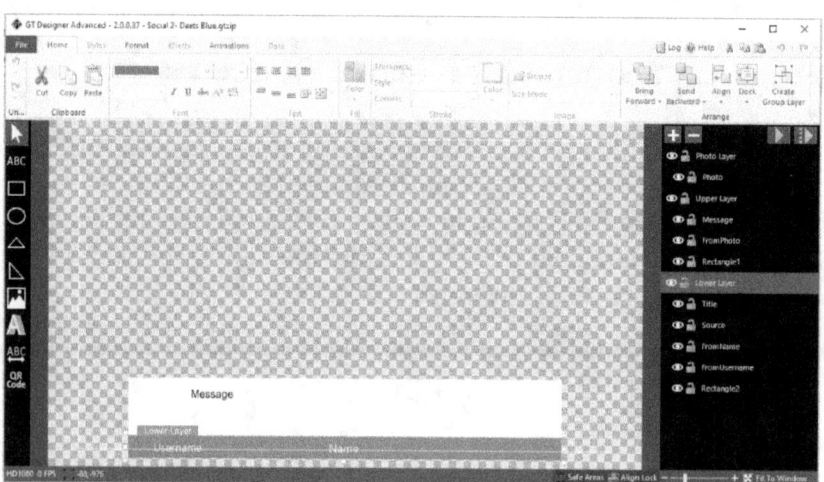

Getting Started

Load the GT Title Designer by locating it in the hamburger menu on the bottom right of your screen. The default resolution for a new design will be 1920x1080. If you want something different, just go to **File** in the upper left corner, click **New**, and you will have the chance to set the resolution.

You will see all the different elements you can use to design your title down the left-hand side of the screen. The images represent text, rectangle, circle, triangle, image, 3-D text, ticker, and QR code. You can select any of the options, then click and drag on the canvas to place it and determine the size.

Once you add an element, you will notice on the right side that it appears under layer one. You can create additional layers by clicking the + button at the top of the layer box and remove them using the -.

Customization Options

Once you have placed an element on the canvas, you have several options for customizing it. For instance, if you place a shape such as a rectangle on the canvas, you can use your mouse to move it around. Hovering over the edges will allow you to drag to resize. Right-clicking on the object will reveal several options, including changing the fill to a different color or image, adding or editing the stroke outline of the object, sending it forward or backward within the layer, aligning it, or docking it to a specific part of the canvas. You can also create a group layer that will allow you to manipulate multiple objects at once and align them with each other.

There are more options available across the top of the editor, including styles that allow you to choose from some preset options. The format tab provides more specific settings for location, dimensions, rotation, and rounding. The effects tab offers options like opacity, shadow, and reflection.

Adding text is like adding a shape. Just select the text option and click and drag to add it to the canvas. Selecting text and clicking on the **Home** tab at the top will bring up text options such as font, font size, color, alignment, and stroke.

Some experimentation with the interface will help you see all the options you have for manipulating shapes and text. Many users will find most of the options like Microsoft Word or PowerPoint.

You can decide if you want this title to hold the actual text you want to display or a placeholder so that it can be edited later. If you wish to use the text as a placeholder to change later, just enter some placeholder text. When you are done, be sure to save your title and exit the designer.

Adding Your New Title

Back in the vMix interface, click on **Add Input** and select title. When the title menu comes up, you can easily find your new creation by clicking on the **Recent** tab at the top. Once you select your title, you will have the chance to replace your placeholder text by right clicking it. Next, you can add that title to one of your overlay channels and activate it to see how it looks with your output source.

This is just a basic look at the vMix GT Title Designer. There is no limit to the ways you can create customized titles to go with your productions.

Using vMix Call

With vMix call, you can invite guests from nearly anywhere to join your live production. All your guest needs is a quality internet connection, the latest version of Google Chrome and a webcam. vMix takes care of everything else. You don't need to worry about extra equipment or common issues like echo or feedback. vMix Call is available with the HD edition with one caller. 4K increases that to 4 callers, and up to eight are available with Pro.

Pro Tip: If your far end guest is using vMix, try using a vMix to vMix connection.

Getting Started with vMix Call

To get started, open your vMix interface and head to the input menu in the lower left. Select **Video Call** at the bottom of the list. You will see two options: Host a Call and Connect a Call. Connect a call allows you to connect to another vMix production by sending your vMix output. But, for now, try hosting a call.

Guest Connection Information

Next, you will see the password box with an automatically generated password for your guest. Your guest can connect simply by going to vmixcall.com and entering their name and that password. You can also send your guest the link in the box below for a direct connection. Just be sure the guest is using Google Chrome on their computer.

Settings

Below that, you will see settings for the return feed. This will control what you guest will see when they log onto the call. You can choose the video source, but it will default to your primary video output which is most likely what you will want your guest to see. If you have the 4K or Pro editions of vMix, you can also choose to send any of the three additional outputs.

Below that, you can choose the bandwidth of the video you send to your guest. This defaults to 720p to 1200 kbps. However, if there are any bandwidth limitations for you or your guest, reducing this will not reduce the bandwidth of your production, just what your guest sees.

Finally, you can choose what audio mix you want to send to your guest. By default, you can send either the master audio or the mix for your headphones. With both, mix-minus is automatically taken care of, meaning your guest will not hear themselves echoing back through the connection. You can also set up custom audio buses to send audio to your guest. However, keep in mind, in those cases, you will need to take care of your own mix-minus.

The box right below that allows you to choose only peer to peer connections. Usually, there is an automatic fallback to a server if you can connect directly to your guest. If you are trying to troubleshoot a direct peer to peer connection, it may be useful to check this box. Now, just click the **OK** button.

The Guest Interface

vMix Call Interface

vMixCall.com

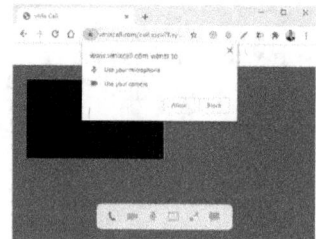

Browser Interface

When your guest connects via the address and password you provided, they will have a reasonably simple interface. It allows them to control what they see and chat with the producer on the other end.

One final tip. You may want to go into the settings for your newly created input and, under general, untick the "Automatically mix audio" checkbox. That way, you can manually control your guest's audio and not have it automatically turn on or off as you move their video into the output window.

Using the vMix Web Controller

The vMix Web Controller gives you access to many of the features and controls of vMix from a smartphone, tablet, or any computer on your network with a web browser. Just about any functioning smartphone or tablet will work if it has a web browser and can connect to the same network as your vMix computer. The Web Controller gives you the option to control the live production while away from the computer or give part of the production responsibilities to another person. The vMix Web Controller is ideal for remotely switching inputs, accessing shortcuts, updating titles, and more.

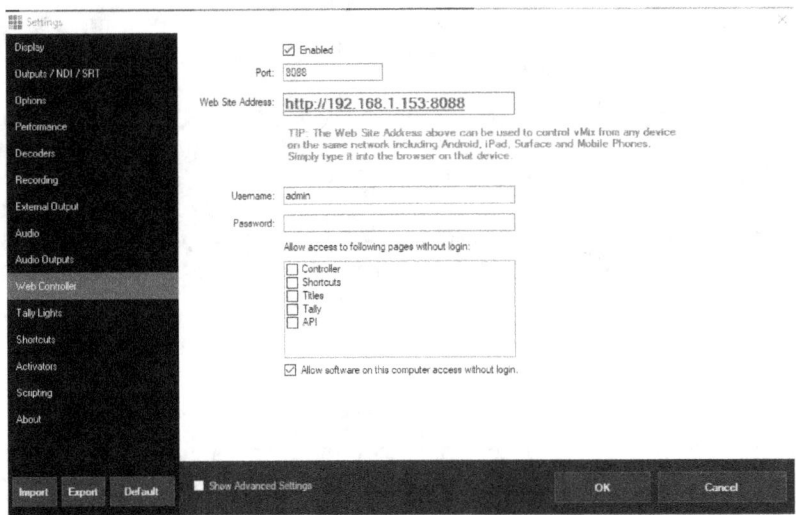

Getting Started

To get started, go to **Settings**, and choose **Web Controller**. First, ensure that the Enabled checkbox is ticked to activate this feature. The **Web Site Address** shown in the "Web Controller" settings area is what you need to access the controller. Copy and paste it into the web browser of the computer you will be using. You can add additional security by assigning a username and password. As long as the password field is blank, no password will be required to log in. If you enter a password, you can give specific access to devices. This is helpful if you want to provide user access to only some features of the controller.

The Interface

Once you enter the web address into the device's browser and log in if necessary, you will see four options represented by icons. Those screens are shortcuts, controller, tally lights, and title editor.

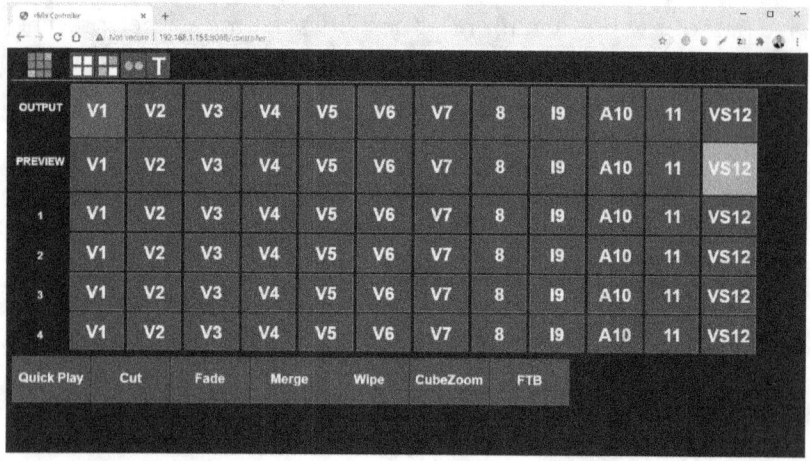

Controller

The next icon, with four white squares, will open the controller page. This gives you a fully functional switcher, just paired down for size. Instead of full images and descriptions of the inputs, each input is designated by number and a letter representing it. For example, a camera on input 1 would be labeled C1, a title on input 2, T2. The top 2 rows represent the Preview and Output windows. The green square will represent what is live in the output window. A yellow square represents what is currently in preview. Across the bottom of the page, you will see all of the standard transitions. Tapping one of those buttons will move the input from preview to output using the selected transition. The next four rows represent the overlays, which can be activated with just a touch.

Web Controller Interfaces

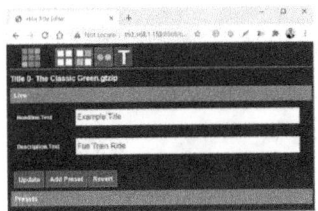

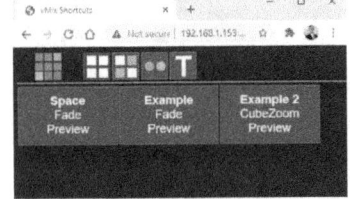

Titles   Shortcuts

Shortcuts

On the shortcuts screen, you will see all the shortcuts you created in the shortcuts editor in vMix. If you wish to limit which shortcuts show up on the web controller, you can tick or untick the Show in Web Controller Box for each shortcut you create. The shortcuts will behave the same way when activated from the web controller as they would from the vMix computer interface.

Tally

The third screen is the tally light screen, which allows you to turn any mobile device into a camera tally light to help talent know which camera is live. By going to that screen and selecting a camera input, the screen will light up yellow when that camera is in preview and green when that camera goes live to the output.

Title Editor

The final page is the title editor. This enables on the fly editing of titles that can update in real-time in vMix. You can select any titles already set up in vMix and change the text. When the title is changed, and the **Update** button is pressed, it will immediately change the title even if it is in the live window. You can also add presets to quickly switch between different text for that title input.

6 vMIX DEEP DIVE

vMix Color Correction Tools

Colour Correction Tools

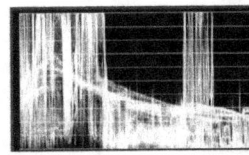  

| Waveform Monitor | Vectorscope | Waveform + Vectorscope + Preview |

A big part of making your live production look great is making sure your colors look realistic and match from camera to camera. vMix includes color correction tools that are advanced enough for color optimization experts yet simple enough that average users can improve their productions' overall look. Color correction tools are available on every video input, including cameras, videos, and images.

Basic Color Adjust Tools

Those who just wish to make basic color adjustments may be content to work with the basic **Colour Adjust** tools. They can be accessed by clicking the gear icon on the selected input and choosing **Colour Adjust**. From this menu, you can adjust the black level, brightness, transparency, and saturation. You can also independently adjust the level of red, green, and blue. An auto white balance feature is also available.

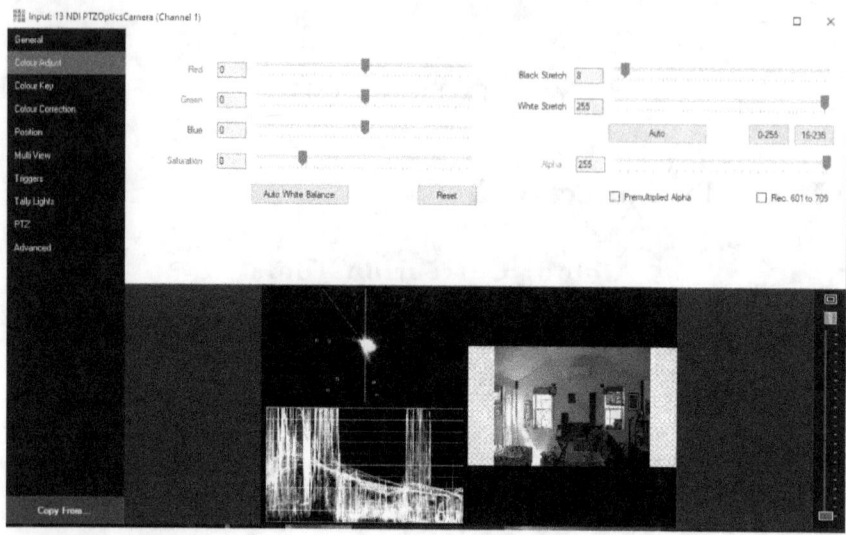

Pro Tip: Most cameras can look better with just a bit of Black Stretch applied to their video feed from the **"Colour Adjust"** area of vMix.

Advanced Color Correction

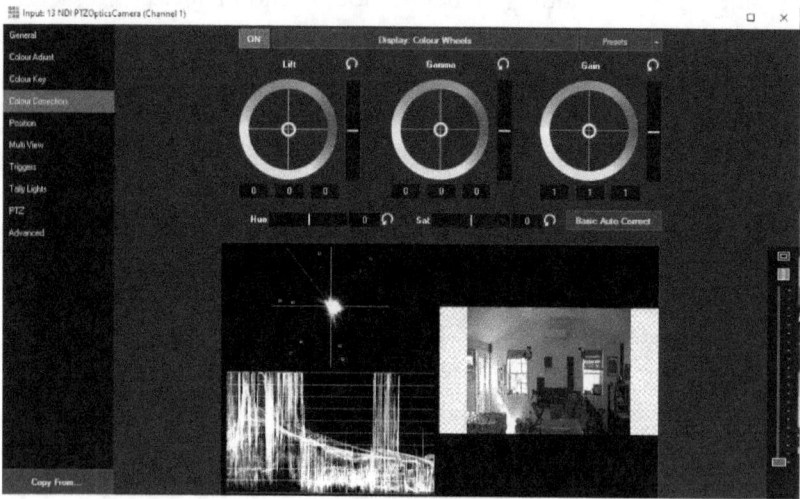

For more color control, select **Colour Correction** from the input settings menu to access the professional color correction tools. The interface offers control over the lift, gamma, and gain. The lift adjusts the dark areas of the image, the gamma impacts the colors in between black and white, and the gain adjusts the bright areas of the image.

The most basic function on this menu is the Basic Auto Correct button that attempts to set the lift and gain controls to their optimum setting. Also found on this page are several reset buttons shaped as a circular arrow to undo any settings that you would like reverted to default.

Color Wheels

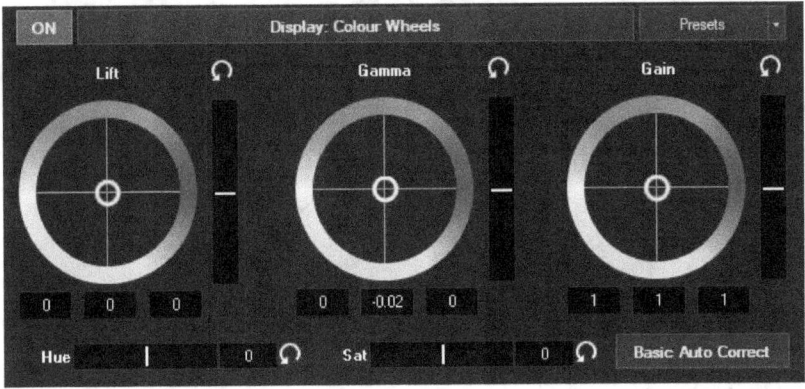

The color correction tools inside vMix have three (3) wheels which are used commonly in video editing software. The first wheel represents the blacks (shadows) also called lift. The second wheel represents Gamma or mid tones and the third wheel represents the highlights and overall brightness of the image. Each wheel adjusts a particular part of the image. The first wheel can be used to work with dark areas, the middle wheel works on areas which are usually skin tones and the final wheel is for the highlights. Using the three components of the color correction wheels allows operators the ability to do fine color correction in an intuitive layout. You can use gain control, for example, to adjust the yellows just in the highlight of an image.

Adjustment can be made on the color wheel by dragging the circle to the desired location. Adjustments can be made to all colors equally by using the Luminance/Brightness bar to the wheel's right. Hue and saturation can be adjusted using the bar below the wheels.

Pro Tip: Try using the ON/OFF buttons to see the color corrections you have applied. Sometimes it's worth seeing all of the changes you have made by looking back the original.

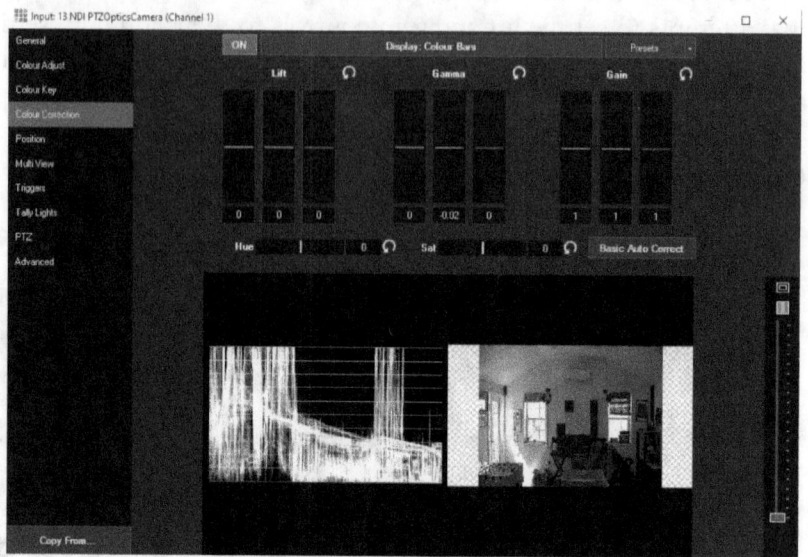

Color Bars

By clicking on **Display Color Bars** at the top, the view will be changed to color bars here. The red, green, and blue levels can be adjusted independently for lift, gamma, and gain. Color bars are easier to operate then wheels because wheels are constantly adjusting multiple colors with every change. Bars allow you to adjust a single color independent of others. If you are just learning how to perform color corrections or you are trying to fix a specific color issue, color bars may be easier for you to operate than wheels.

What is a Vectorscope?

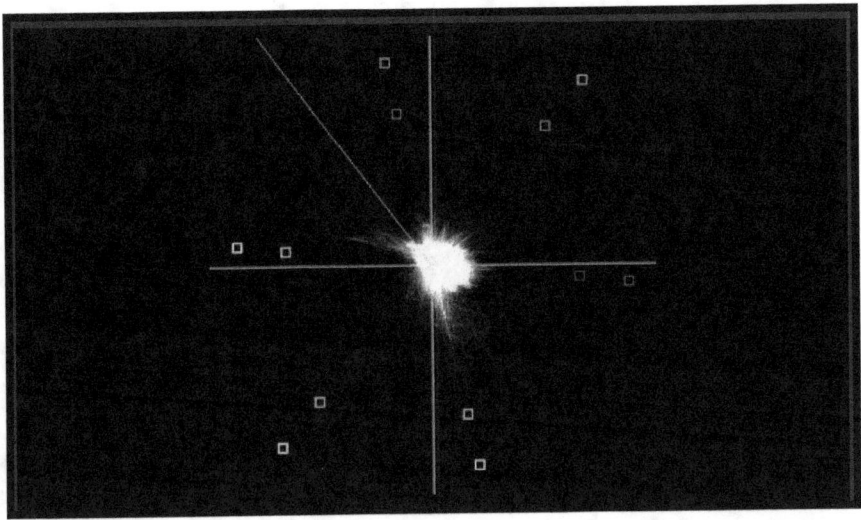

A Vectorscope is a tool that represents the color of your image. It is an x and y graph representation of the color accuracy of your live video feed. At the top of a Vectorscope you have red, toward the bottom you have cyan and there is also a green. The Vectorscope graph allows you to see the balance of the colors coming from your live video. To accurately tune a camera, you can put up a color chart in your space and zoom into it with your camera. The vMix color chart will produce lines that connect your current image with the ideal colors. A Vectorscope is a tool that is ideal for live video color correction without having to rely on your own eyes and potentially inaccurate monitor representations of an image. The Vectorscope is all about color and provides you with tools to accurately adjust your camera settings ideally on the camera side first. Once the camera has been accurately tuned the finishing color corrections can be done in vMix.

What is a Waveform Monitor?

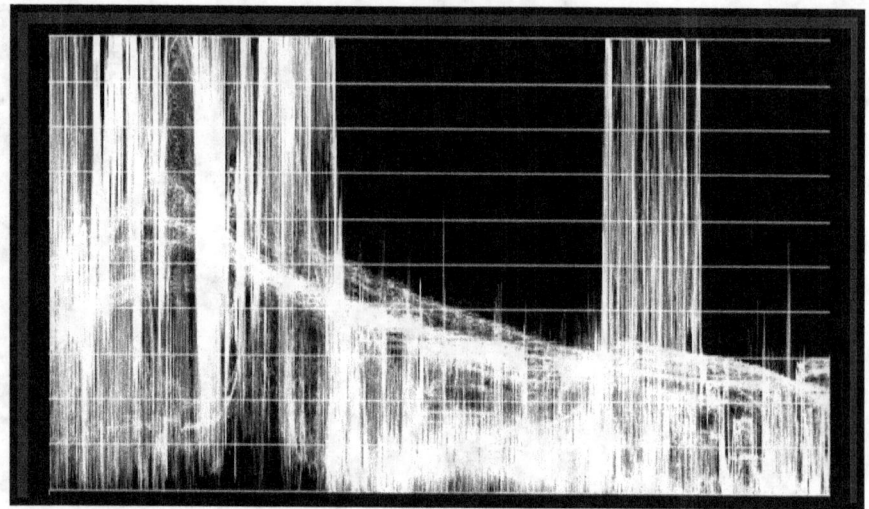

For those with color correction experience or those who want to learn, vMix also includes several types of Waveform Monitors. The Waveform Monitor is the counterpart for the Vectorscope available to handle brightness and exposure. With a Waveform Monitor, you can easily see if your image is clipping at the top or if the blacks are getting crushed. The Waveform Monitor will allow camera operators the ability to adjust the image preferably in the camera first to ensure your image has a good exposure.

Ideally, you want your camera's image to be within the limits of your Waveform Monitor. You can use vMix to make sure your video has perfect whites and blacks using the tools as needed. The tool can be accessed by clicking the small color bar box on the right side of the video preview. Each of the waveform monitor options can be viewed in a split-screen with a preview of the input or in a full screen independently. These are essential for completing precise color correction adjustments. Using the slide on the right, the zoom level of the image can be adjusted. Users can make adjustments to the color setting wheels and bars while seeing the precise impact on the waveform monitors. This can often help even novices detect and correct color issues.

Color Correction Presets

Another tool built into the vMix color correction system allows users to save preset files that can be uploaded to other vMix inputs. This preset information can be useful for color matching cameras from different manufacturers. The preset will save your color correction information which can be applied to any other input in vMix. These presets may not get your image 100% corrected because every camera's exposure will be slightly different.

Pro Tip: One of the key things with color corrections is starting with exposure and color settings inside the camera. Many times the highlights can be blown out before they even get into your video production software. Using a Vectorscope and waveform monitor you can bring down the range of your camera to make sure that you are not clipping your sensor's full potential. These tools help you make sure that you are not crushing the blacks or overexposing the whites with the settings in your camera. The waveform monitor specifically allows users to stretch the exposure to the perfect white and black settings.

Input Video Preview Options

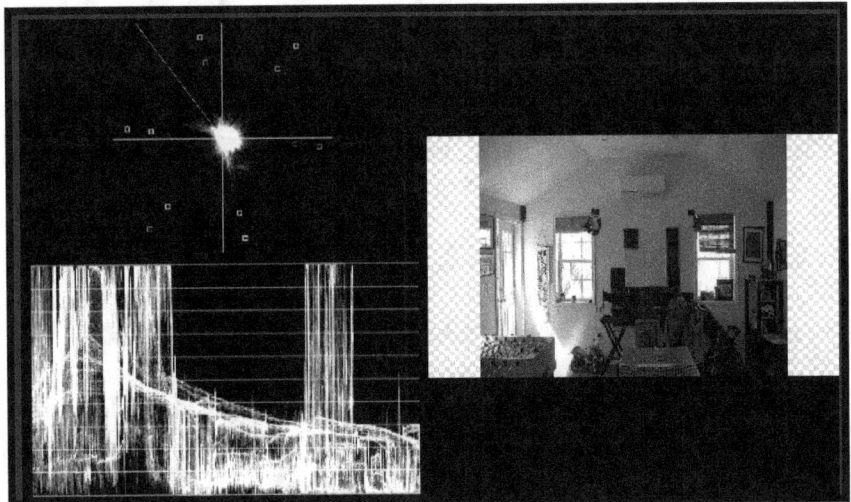

Underneath the color correction tools you will see the live video preview of your input. On the right side of this interface you will see options for changing this video to see the Waveform Monitor, Vectorscope and various combinations that include a live video preview. You can also apply video standard grids for SMPTE, square and vertical video for monitoring.

How to use Virtual Sets in vMix

With built in virtual sets, vMix makes it possible to create a professional looking set for your production with nothing more than a green screen and some decent lighting. Virtual sets are easy to use and they can look great. You can even have a custom virtual set built specifically for your needs.

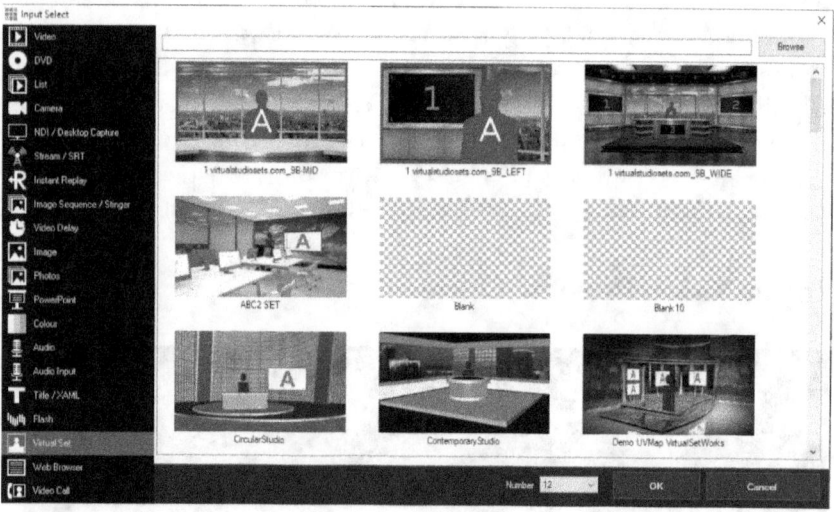

Getting Started

To start out, just make sure you have a green screen behind you or whoever your talent is for the production. Green screens can be purchased online in fixed and portable formats. You can even create a permanent solution by painting a wall. You will need to get your lighting right so that your talent and green screen are well lit. This will allow the keying feature to remove the entire background appropriately.

Color Key

The first step is to set up an input with the camera you wish to use for the talent. Then, go into your camera input settings and choose **Color Key** from the options on the left side. Use the eyedropper to select a sample of the green screen. You can adjust the presets to get it perfect or set one of the presets that will often give you excellent results. Close out of that menu and you should now have a camera input with your talent and no background. Now you are ready to add a virtual set input.

Pro Tip: If your green screen does not cover the entire picture that your camera is capturing, use the crop feature to remove unwanted areas on the sides.

Adding a Virtual Set Input

Click on **Add Input** and select **Virtual Set** from the options on the side. Now you get to choose your virtual set. Click **OK** and the virtual set will show up in your inputs.

Customizing Your Set

Virtual Set Input

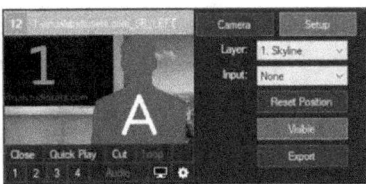

Virtual Set Input   Virtual Set Setup

Now it is time to start customizing. Next to your virtual set input you will see your customization options. When you select "Camera" you will have access to the various zoom options for your set. When you click "Setup" you can access all the various layers in the set. Those may include the backdrop, virtual screen, talent, and any other customizable parts of your set that can be assigned to an input. Select the layer and then assign the input you wish to use. For instance, for the talent layer, you will want to assign the camera you just set up with a green screen and color key.

With the layer selected, you can also customize the position of the input in the set. For example, to get the talent in the right position, just click and drag in the preview window and to resize just hold down shift while dragging. Continue to assign layers and make adjustments until you have completed all the features of the set you want to use.

Once you understand how the virtual sets work, there are multiple set options, the ability to create your own from scratch, and the option to import custom sets from third-party sellers. You can also use the export button at the bottom of the virtual set input to save all of your settings to use the set later or import to another instance of vMix.

Note: The online course available on Udemy includes over 15 additional virtual sets that you can use for vMix. These virtual sets have been designed for sales, marketing, houses of worship, finance, and educational purposes.

Working with NDI Sources in vMix

NDI or Network Device Interface is a powerful way to share video content across a local area network. The protocol, developed by NewTek, opens new production possibilities, and with vMix, it is easy to both send and receive video via NDI. Not only can vMix share cameras, video, audio, and graphics to other vMix devices, it can both send and receive from any devices or software that support the NDI protocol.

NDI stands for Network Device Interface and it is a high quality, video-over-IP standard developed by NewTek to enable video-compatible products to communicate, deliver, and receive high-definition video over a computer network ideal for live video production.

What is NDI used for?

Many video projects use NDI to send and receive video over IP. NDI features an auto-discovery feature which makes managing video sources available on a network very easy. For example, a church may use NDI to send PowerPoint slides from one computer and receive them on another computer used for live streaming. Another example would be setting up a display in an office to show an NDI source coming from a video production software like OBS. Any Windows or Mac computer can receive the NDI video stream and display it on a TV located in a facility using free tools such as the NDI Studio Monitor.

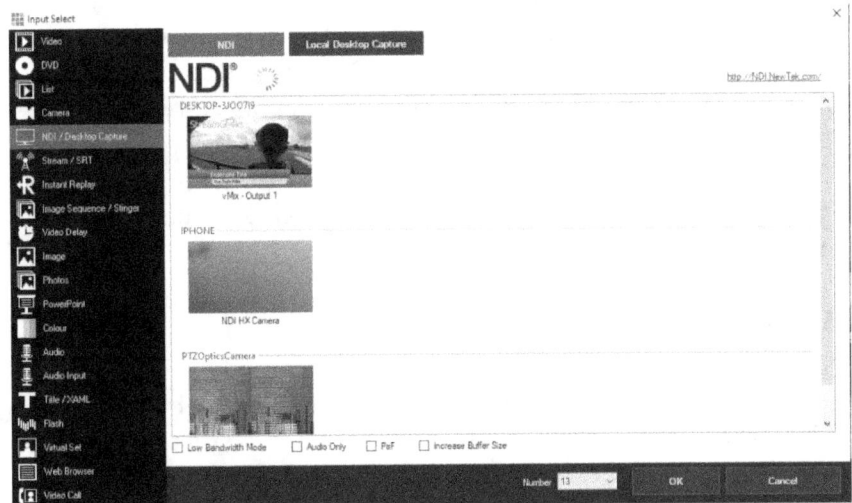

Adding the Source

First, be sure that your vMix computer and your NDI source are connected to the same network. Adding an NDI source in vMix is just as simple as adding any other source. Just go to **Add Input** and select **NDI/Desktop Capture**. You will see all of the computers and devices on your network that are connected via NDI. Under each computer or device, you will see any available NDI sources. At the bottom, you will see a few checkboxes. One will select only the audio from the source. The other will receive the source in low bandwidth mode. Just click on a source to add it as an input. Once you have the input setup, you can change it to another NDI source by just right clicking on the input and selecting a new source.

Using the NDI Source

Once the source is ready, you can use it just as you would any other video input. You can toggle the audio on and off, preview it, or send it to an overlay channel. By clicking on the gear icon, you can access the full settings menu. When you open the settings, you will see that the NDI source settings are very similar to the settings on a directly connected camera and can be adjusted the same way. You can change the name for easier reference, set the mouse click behavior, mirror it, sharpen it, and enable or disable the automatic audio mix. You also have access to color correction and the ability to set up a chroma key if the source is set up with a green screen. You can make position adjustments, add it to a multiview, and even set up triggers and tally lights.

Pro Tip: Unlike normal inputs, you can right click inputs to open up additional options.

Bandwidth Matters

One important thing to consider with NDI is network bandwidth. While NDI sources are local and do not require internet bandwidth, they still need sufficient internal network bandwidth. For the best results, be sure you are using a gigabit network.

Note: NDI stands for "Network Device Interface" and NDI® HX is the "High Efficiency" lower bandwidth version of NDI. NDI HB stands for the "High Bandwidth" version of NDI. NDI features a combination of high-quality video with low latency transmission that is ideal for live video production. NDI|HX is ideal for broadcast professionals adding NDI sources to an existing network that has not been originally designed for video production. NDI|HX video sources are generally one tenth of the bandwidth of full NDI sources. See the bandwidth comparison chart.

NDI Mode	Bandwidth

HDI HX Low (720p60fps)	6-8 Mbps
NDI|HX Medium (1080p30fps)	8-12 Mbps
NDI|HX High (1080p60fps)	12-20 Mbps
NDI|HB (1080p60fps)	125-200 Mbps (Nominal Range)

Example:

NDI Device Examples (1080p60fps)	Bandwidth	Accumulated Bandwidth	Total % of Gigabit Network Switch
NDI Scan Converter on Laptop for Powerpoint slides	125 Mbps	125 Mbps	12.5%
2 x NDI Monitor for camera operators	125 Mbps / Each	375 Mbps	12.5% / Each
vMix System output in 1080p60fps	125 Mbps	500 Mbps	12.5%
NDI Monitor in Overflow Room	125 Mbps	625 Mbps	12.5%
5 x PTZOptics NDI|HX (High)	12 Mbps / Each	685 Mbps	1.2% / Each
Suggested Headroom	250 Mbps	910 Mbps	25%

Total Usage			91%

Tip: Network Bandwidth head room recommendations can vary widely from 30% - 60% depending on what the network is utilized for. Please consult your network administrator before adding NDI sources to your local area network. Newtek suggests "NDI traffic should not take up more than 75% of the bandwidth of any network link.".

Sending NDI from vMix

Keep in mind that it is also simple to send NDI across your network from inside vMix. Just go to your **Settings** menu and select **Output**. You can then easily assign any of your output to send an NDI signal from your Output, Preview, MultiView, or any input source.

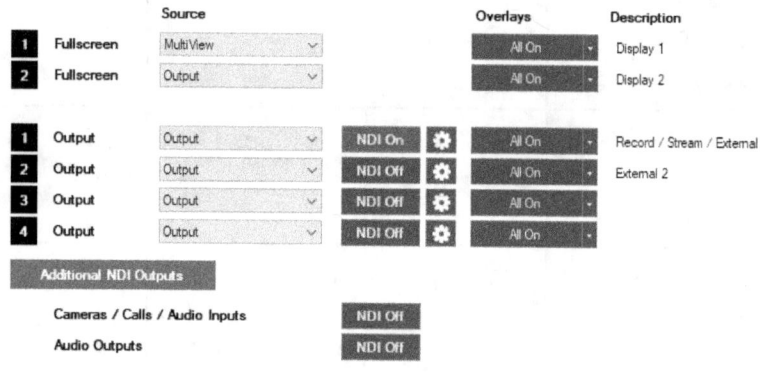

What is an NDI Camera?

NDI cameras are able to communicate using the Network Device Interface or NDI protocol. They can connect to a LAN (Local Area Network) and seamlessly integrate with hundreds of software applications including OBS, Wirecast, vMix, xSplit, NDI Studio Monitor and much more.

What is an NDI camera used for?

NDI cameras often have PTZ (Pan, Tilt and Zoom) functionality which takes advantage of the two-way communication capabilities of NDI. In this way, NDI cameras can be controlled over the same single ethernet cable used to send audio and video. For example, a PTZOptics NDI camera can use a single ethernet cable to power the camera, control the PTZ functionality of the camera, and to send audio and video to a source on the network. PTZ camera controls inside of vMix will be reviewed in an upcoming chapter.

What is the difference between NDI and SDI?

SDI is a technology that has been around for decades. SDI stands for Serial Digital Interface, and the cable itself is capable of sending uncompressed video long distances. NDI is a much newer technology that uses the latest video compression methods to make sending and receiving high-quality video possible over standard computer networks. An SDI camera video feed can be converted into an NDI stream and sent over the network. An NDI video feed can also be converted into an SDI video output and plugged into a monitor.

How do I set up an NDI camera?

Most NDI cameras are plug and play when it comes to setup. NDI cameras can be plugged into any LAN (Local Area Network) and configured to operate with any software or hardware solution that supports NDI. Once an NDI camera is plugged into the network, it will show up as an available source on your network; therefore, the friendly NDI name that you give your camera will show up in any software or hardware solution when you click the "add NDI source" option.

How can I capture video from another computer?

Perhaps one of the most popular ways to get started with NDI is sending video from one computer to another. Popular examples of this include sending PowerPoint slides from one computer on your local area network to the main vMix video production software. vMix has a handy tool called vMix Desktop Capture that you can download for free to enable a secondary Windows computers to send video via NDI back to your main vMix computer.

vMix Desktop Capture for NDI is incredibly easy to use. Once you open it up, it will search your computer for potential screens and applications to capture and it makes them available on your local area network via NDI. Back at your main vMix computer you will instantly see all of the available screens and windows available to connect with via NDI using the vMix Desktop Capture for NDI tool.

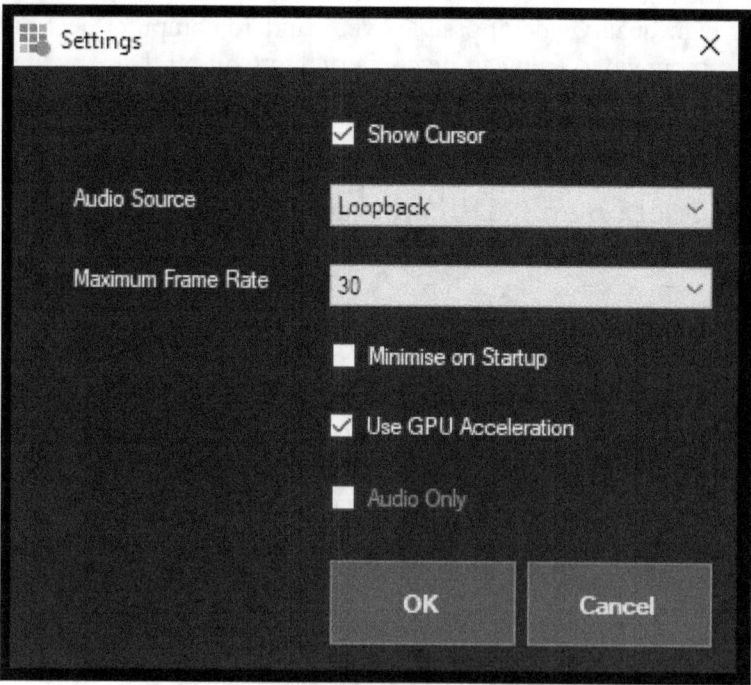

There are just a few options that are revealed when you click the **Settings** button. vMix Desktop Capture for NDI allows you to show or hide the mouse cursor. You have the option send loopback audio, choose silence or a specific audio interface. You can also minimize this application at start up and enable GPU acceleration. It is that easy to use!

PTZ Camera Controls in vMix

PTZ Camera Controls

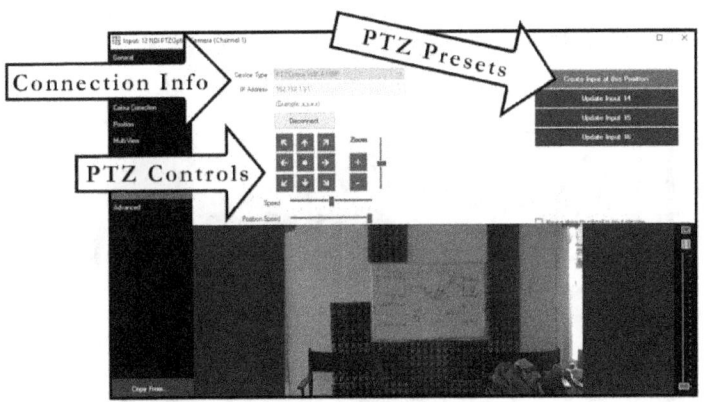

vMix has modernized live video production and increased what small video production teams are able to do. vMix 4K and Pro editions include multiple ways to control pan, tilt and zoom cameras which can help automate video production workflows even further. vMix can currently control specific PTZ cameras over an IP connection and USB cameras that support UVC.

Here is the list of PTZ cameras that vMix currently supports:

1. Sony
2. Panasonic
3. PTZOptics
4. iSmart
5. HuddleCamHD (via USB)

Enabling PTZ Controls

You can enable PTZ camera controls inside of vMix by opening an input and clicking on the **PTZ** tab. Here you can select one of the supported devices in the dropdown menu. Most PTZ cameras will require a static IP address for vMix to connect. USB cameras like the HuddleCamHD Pro can be controlled via a USB connection by selecting the **UVC PTZ** option.

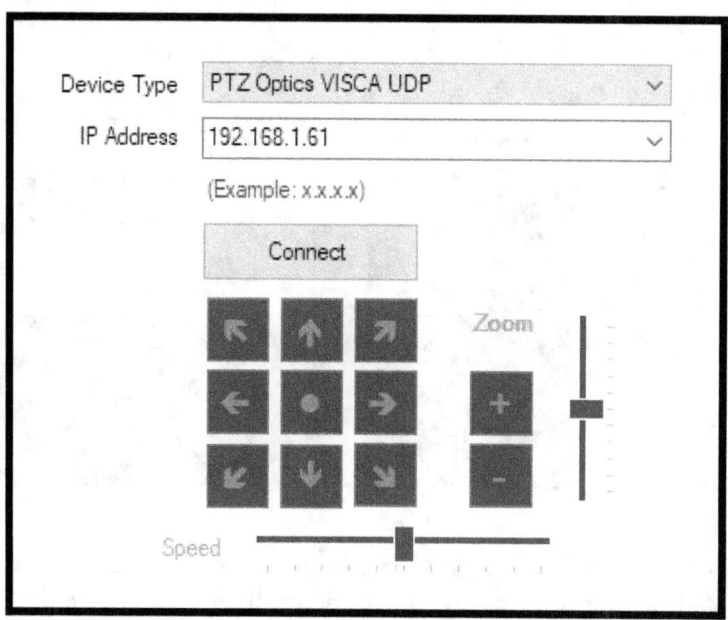

Ethernet Controls

To gain IP controls of any IP supported camera you must make sure your computer is connected to the same network as your PTZ camera. You will also need to know your camera's IP address. PTZOptics cameras for example can receive an IP address in two ways. One is a statically assigned IP address, where you manually give the camera an IP address on your network. The other way is via DHCP. DHCP allows your camera to automatically receive an IP address from your network's router.

Pro Tip: Your camera and computer should be on the same IP range. You can find the IP address of your computer using the **Command Prompt** in Windows and typing in "**ipconfig**" and pressing **Enter**. Look for your computers IPv4 address. Make sure your camera's IP address is on the same IP range which means the first 3 sets of numbers are the same, and the last number is unique.

Click **Add Input** and select **Camera**. If you have an NDI camera click **NDI** and choose your camera from the list. Select your camera from the list and click **OK**. Once your camera is inside of vMix you can click the settings cog to find the **PTZ** section. Inside the **PTZ** section you will see a device type dropdown menu.

Note: ZCam camera models from PTZOptics do not offer pan and tilt but vMix can control presets and zoom.

Once you select your camera's control method, you can enter your camera's IP address and press connect. This you will unlock the pan, tilt and zoom controls of your PTZ camera. You will notice two control bars which allow you to select the speed of your pan, tilt and zoom. There is also a slider designated to control the speed at which the camera moves in-between presets. This is ideal for adjusting the camera to perform slow moving scenes that sweep across an area, or quick movements that are hidden from viewers while the camera is in preview.

You should also notice a **Create Input at this Position** button at the top right of the input settings window. You can click this button to create a PTZ preset inside of vMix. When you click this button vMix will create a new input inside of your vMix production. When you bring this input into preview the camera will automatically recall the camera's PTZ preset position.

You will notice that vMix takes a screenshot of the camera's position when saving the preset inside of vMix. In this way you recall PTZ presets with a visual aid that is inside each thumbnail. The idea for professional productions is that when you use one of the PTZ presets in the preview window the location of the PTZ preset moves. In this way, you can hide the PTZ movements from your audience if you would like to.

It is worth noting that PTZ camera controls are also available inside vMix shortcuts. Therefore, you can control PTZ cameras using a variety of hardware solutions that integrate with vMix. For example. you can use an Elgato StreamDeck, an Xkeys controller, or even a USB connected xBox controller.

UVC Controls

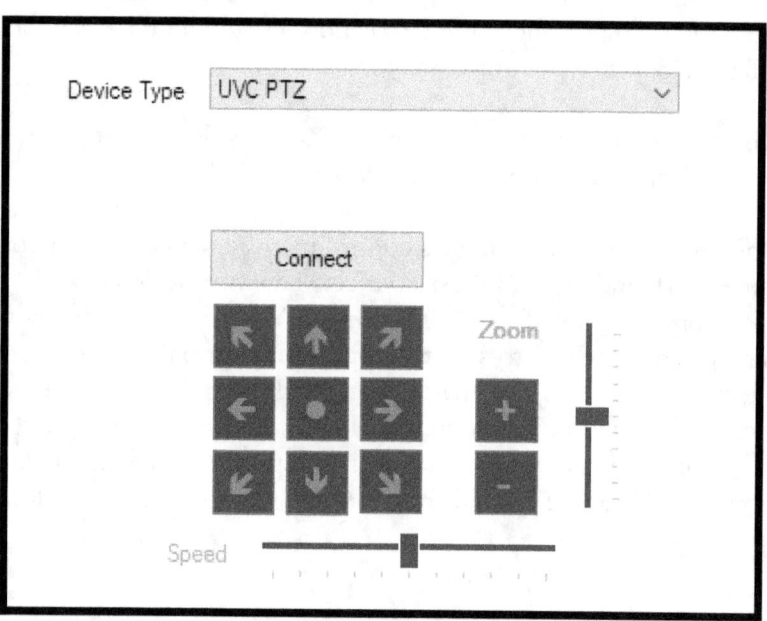

UVC controls allow you to control webcams that support UVC. To use this PTZ method, simply bring your camera into vMix using the camera input via USB. Then inside of the settings window, in the PTZ section choose **UVC PTZ** and click connect. Wow. It's that easy. Now you can operate PTZ camera control functions inside of any PTZOptics or HuddleCamHD connected USB camera. UVC controls are limited in comparison to network connected PTZ cameras but they are easier to connect with plug and play USB cabling. The HuddleCamHD Pro is an interesting camera that was designed to be used like this. USB cameras with high resolutions sensors like the 4K HuddleCamHD Pro can be used with UVC or virtual PTZ options inside of vMix very effectively.

Digital PTZ controls

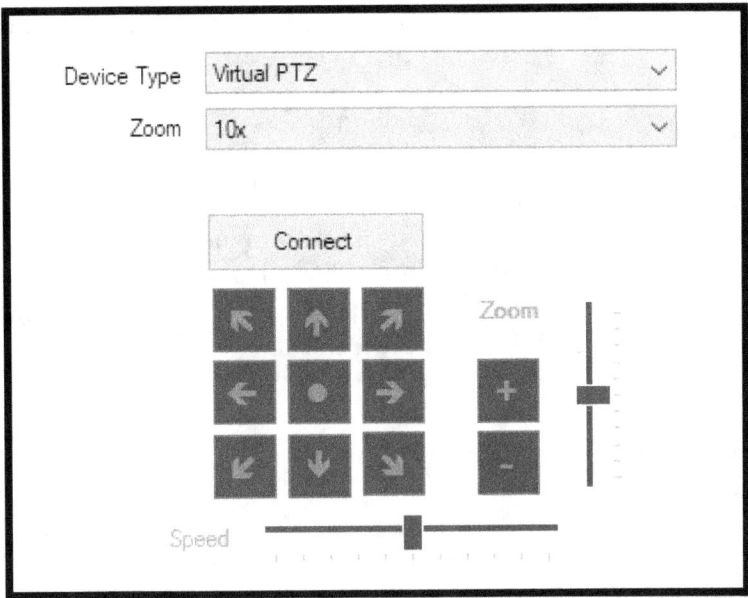

Finally, vMix features virtual PTZ options. High resolution cameras like the HuddleCamHD Pro allow you to digital zoom in using 720p or 1080p productions without loss of quality. You can also use cameras with 4K Blackmagic capture cards to bring the video into vMix and operate with digital PTZ controls.

Once this input is inside of vMix, in the settings window, in the **PTZ** tab, you can select **Virtual PTZ** and choose a **Zoom Limit**. Once you press connect you can now operate the digital pan, tilt and zoom inside of vMix with the set Zoom Limit. You can set presets and operate the input like a real PTZ camera.

Pro Tip: Try the merge function between two digital presets. It almost looks like a real PTZ camera! Digital PTZ controls can be added to almost any input.

External Output (Connecting with Zoom)

Seemingly overnight, Zoom became the go to video conferencing platform for businesses, non-profits, government agencies, and schools. While Zoom is a powerful product in its own right, users quickly began to look for ways to integrate other video tools. One of those tools is vMix. Users want to be able to feed the live video produced in vMix to the Zoom software and integrate the video from Zoom into their live video production in vMix. Fortunately, thanks to the power and flexibility of vMix, this is not a problem.

There are multiple ways to do this, but the simplest method is to simply use vMix to output video directly into Zoom using the virtual vMix webcam and the **External button**. The explanation below shows you how to use two computers on the same network and sharing the audio and video via NewTek's NDI protocol. This is a great introduction to the power of NDI and it's use for video production.

Here is what you need to get started:

- A separate PC laptop or desktop connected to the same network as your vMix computer.
- The vMix Desktop Capture program available at https://www.vmix.com/software/download.aspx#downloaddesktopcapture running on the second computer
- The Virtual Input program in the Newtek NDI Tools Pack: https://ndi.tv/tools/ running on the second computer.

Here is how to set it up:

1. In the vMix settings menu, choose the Outputs/NDI tab and turn on the NDI button next to Output 1.
2. Click on the gear icon next to that output and select Bus A from the Audio Channels dropdown.
3. Go to the audio outputs tab in settings and change the Bus A setting to enabled.
4. Go to the audio mixer. Activate the Bus A button for any audio inputs you want to be sent to Zoom. If you are going to share your entire live production, that would likely be all your used audio input sources.
5. On the other PC, open the NewTekNDI Virtual Input software you downloaded as part of the NDI tools pack. To get to the options, click on the up arrow inyour Windows task tray's bottom right and look for the yellow NDI icon. Click on it.

Click on the name of the main vMix computer and then select Output 1.

6. Open Zoom on your second computer and start your meeting.
7. Click on the up arrow next to the microphone in the lower-left corner and select, as your microphone Line (NewTek NDI Audio)
8. Click on the up arrow next to the video camera and select Newtek NDI Video as your camera.
9. Run the Desktop Capture software on the second computer.
10. To bring Zoom into vMix, use the Add Input button and select the NDI/Desktop Capture option. Choose NDI, locate, and click on the computer and display where Zoom is currently showing.

You are ready to go! One quick warning: Be sure not to accidentally enable the Bus A audio button on your desktop capture input. Otherwise, your guest will hear an echo.

Using vMix Playlists

A great way to automate your productions in vMix is by using the playlist feature. This feature allows you to create a playlist of inputs and automatically transition between camera, videos, and other inputs you have created.

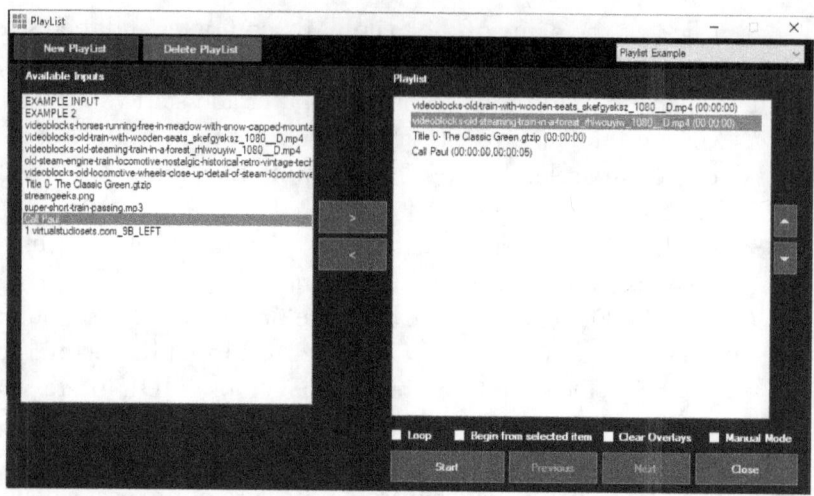

Getting Started

Once you have your input sources set up in vMix, click on the gear icon next to the **PlayList** button at the bottom of the interface. This will bring up the **PlayList** dialog box. Click **New PlayList** and give it a name. If you have an existing playlist that you would like to work from, you can check the Copy Current PlayList box. Once you have multiple playlists, you can select the one you want to work with from the dropdown menu in the dialog box's upper right corner. Keep in mind that playlists are connected to presets. So, if you want to save your playlists, be sure to save your preset.

Adding to the PlayList

Adding inputs to the playlist is as simple as clicking on the available inputs in the left window and then clicking the right arrow. Remove them from the playlist by clicking on the input in the right window and click the left arrow. Once you have multiple inputs on the playlist, you can select any one of them and move them up and down in the list.

Adjusting Input Settings

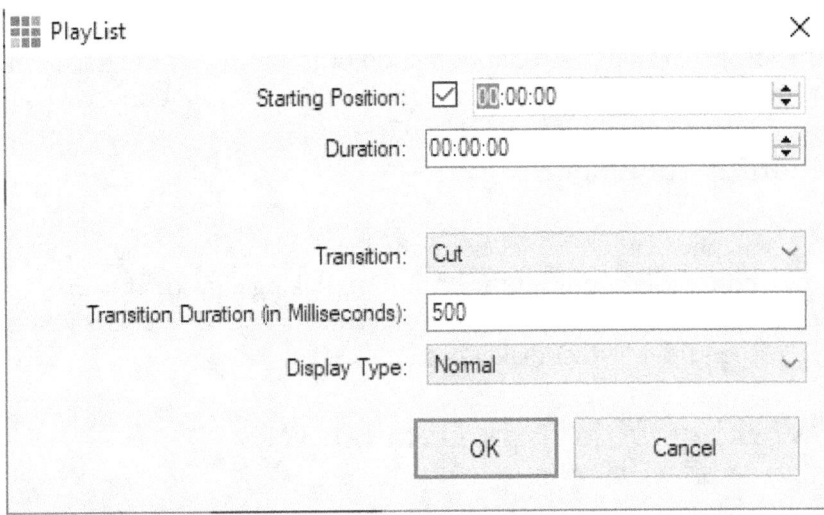

Once you have added your playlist items, you can click on them to adjust their individual settings. This will open another dialog box. For videos, you can set the starting position to the point in the video you want it to begin. You can also set the duration. For videos, this would set how long the video would play after the starting point. For cameras or other sources, this determines how long the playlist will stay on this input before moving to the next item. You can set the transition type to any of the available vMix transitions and set that transition length. Finally, the display type will determine whether the input will be shown in full screen or assigned to one of the pre-configured overlay channels. Now we can click **OK** and repeat this process for each of our inputs.

Pro Tip: If you choose inputs with PTZ camera presets, you can effectively automatic PTZ camera operations. In this way, you can have cameras move to various locations with PTZ presets that are called in a looping playlist.

Changing PlayList Settings

When you are done configuring our inputs, we have a few more options in the PlayList dialog box. At the bottom, you can choose if you want the playlist to loop. Ticking the **Begin from Selected** item box will begin the playlist from any selected input instead of from the beginning of the list. Clear overlays will clear any overlays on the inputs between transitions. Finally, manual mode will keep the playlist from advancing from input to input instead only advancing when the previous and next buttons are pressed. When you are done, click close.

Starting the Playlist

To start the playlist, just click on the PlayList button at the bottom of the interface. If you want the option of making adjustments or manually advancing the playlist, you can simply click the gear icon again to reopen the PlayList dialog box.

Working with vMix Data Sources

vMix offers several options to add dynamic content to your live productions. One of them is vMix Data Sources. With Data Sources, you can import data from sources like spreadsheets, RSS, JSON, a text file, and XML. Even better, this data can be updated in real-time to your live production.

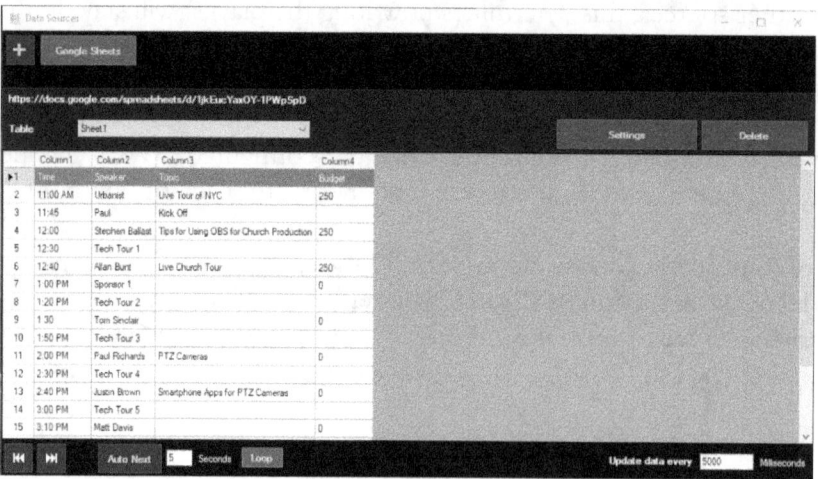

Getting Started

To get started, find the hamburger icon on the lower right of the vMix interface and click on **Data Sources Manager**. Clicking on the plus icon in the upper left corner of the dialog box will reveal the different data sources you can use in vMix. Clicking on a source type will open up a dialog box with the information for that source. For local files like Excel or CSV spreadsheets or text files, you will have the opportunity to browse for the file on your computer or network. For online sources, the dialog box will ask for a URL and any other information necessary for access.

Once you have selected your data source, entered the necessary information, and click **OK**, the onscreen table will begin to populate. With data in your table you can assign that data to a title to bring it into our production.

One great example for using Data Sources is Google Sheets. You can have someone manage a Google Sheet table with various fields of data that you need updated. You can then integrate Data Sources with Google Sheets and pull that data directly into your vMix titles.

Adding a Title Input for Your Data

Go to **Add Input** and select **Titl**e from the list on the left. Choose Title from the tabs across the top and select a title that works for your information. Optionally you can create custom titles using the **GT Title Editor**. Then click OK. This will bring up another dialog box where you can link your data to this title. In a simple title, there will be fields for headline and description. Starting with the headline, click on **Data Source** at the top of the box. In this dialog box, choose the type of data source you are using. If you have already set it up in the data source manager, the table name will be brought in automatically.

Depending on the type of data you are using, you may want to leave the column and row settings as they are or customize them for your needs. By default, this will initially pull the first column and row. When you are done, click **OK**. You can choose which row of data you would like your titles to use via Data Sources by clicking on the row. You can also choose to have the data loop through the rows.

Now go and do the same for the description. Leave the column on auto if you want the second column to appear since the first was used for the headline. You can also manually select the column you wish to use. When you are done, click **OK**.

Formatting Your Title

Your title will now appear in the input section. You can click on it to see it in the preview window with your data. With the title editor still open, you can make format changes to the text in your title. Using the buttons at the top, you can change the font and adjust the size and color.

Additional Settings

There are a couple more settings within the data sources manager that will impact the data's behavior on your title input. Go back to the data sources manager. At the bottom of that box on the left is a button for Auto Next. Click this and set the time if you would like it to move down the rows at set intervals updating the data. You can also set it to loop if you would like it to start over when it is done.

On the bottom right, you can select the update interval. This is useful if the data from your source is being regularly updated. For instance, if someone is entering new data into the spreadsheet you are accessing, this number will determine how often it checks for changes.

7 MASTERING VMIX

vMix Instant Replay

The vMix instant replay system is quite powerful, and it is available for one camera in the 4K version and 4 cameras in the Pro version. The instant replay system essentially records video from cameras to the hard drive so that the video footage is available for replay instantly. vMix's instant replay system also supports slow motion playback which is ideal for high frame rate cameras.

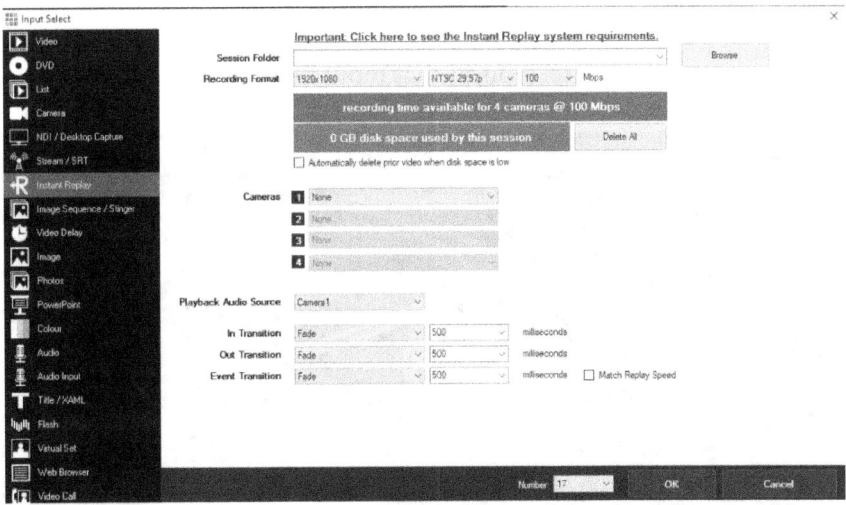

The vMix instant replay system works best when recording at 100 Mbps on a solid-state hard drive. That is 100 megabits per second for each camera. So be very careful not to record instant replay video footage for too long or you will run out of hard drive space quickly.

If you plan on using your instant replay footage in slow motion, frame rate becomes very important to preserve the quality of your replay video. For example, if you broadcast in 1920x1080p video resolution at 30 frames per second, your main live stream being broadcast has 30fps. If you set your cameras and instant replay system for 60 frames per second you can slow the video down by 50% and still deliver smooth

How to use vMix Triggers

vMix Triggers can add powerful automation sequences to your live productions. This becomes especially important for complex productions with a single operator. The producer will not need to remember every single cue and function since many can be automated through triggers.

Triggers can handle anything from adding overlays after a transition to multiple-step processes that will run much of a live show for you. Some examples of common uses include a trigger that starts recording and turns on the microphones at the end of a countdown sequence. Another could be used for switching to the primary camera when a video ends. There is really no limit to what you can automate with triggers.

Triggers are very similar to shortcuts. They both start with a function. With a trigger, it is the press of a key or button that begins the function. With a trigger, the function is started automatically when some condition is met. Shortcuts and triggers can be used together with a shortcut starting a sequence that continues automatically through triggers.

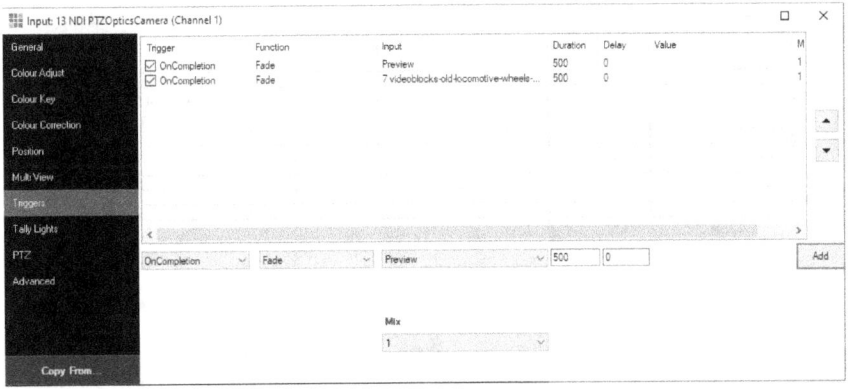

Setting Up a Trigger

Setting up a trigger in vMix is straightforward, and it is done through the input settings. Select the input you would like to work within the input section and click on the gear icon in on the lower right. Once the settings menu for that input is open, click on Triggers on the left column. If you have any triggers already set up, they will be listed here.

Under the trigger list are the fields for setting up a new trigger. The first column is the actual trigger which is the event that starts the process. If you click the dropdown menu, you will see that triggers can include OnCompletion, OnTransitionIn, OnTransitionOut, OnOverlayIn, OnOverlayOut, OnCountDownCompleted, and three audio triggers.

OnCompletion is perfect for triggering a function when something, like a video, is finished. The transition triggers are based on fading or cutting to an input. The overlay triggers work when an overlay is activated or turned off. With the countdown trigger, you can attach a function to the completion of one of the built-in countdown timers. The audio triggers are relatively basic and can trigger a function when an input audio source crosses a dB threshold.

The second column lets you set the function that will occur for that input when the trigger occurs. You can choose from hundreds of functions by clicking on the dropdown menu and selecting the category on the left side. In the next column, you can choose the source used for the function. For instance, if the trigger is set to transition from your input source, this is where you would choose what it should transition to. If you are using any sort of transition function, the next column lets you set the duration. The next column allows you to set the delay before the function takes place.

You can set up multiple triggers on each input and easily change the order in which you wish them to happen.

There is one crucial thing to keep in mind if you are using MultiView of layered inputs. Even if triggers have been enabled, they will only work on the primary layer. That means if you create an input with triggers and then include that input in a MultiView input, those triggers will not function. You will need to add triggers to the MultiView input itself.

LIVE LAN Video Distribution

LIVE LAN is a video streaming option available in vMix which is designed for distributing video over a local area network (LAN). This feature uses HTTP Live Streaming (HLS) which is a streaming protocol from Apple. HLS is an ideal encoding and streaming protocol for use on a local area network (LAN) because it requires low amounts of bandwidth and it's easy to connect to from a variety of receiving devices. One main benefit of HLS is that it sends video using HTTP which can be used by any web browser that supports HTML5. As of 2022, approximately 75% of web browsers already support HTML5 making HLS easy to use with almost all modern web browsers including Google Chrome, Firefox, Internet Explorer, Microsoft Edge, and Opera.

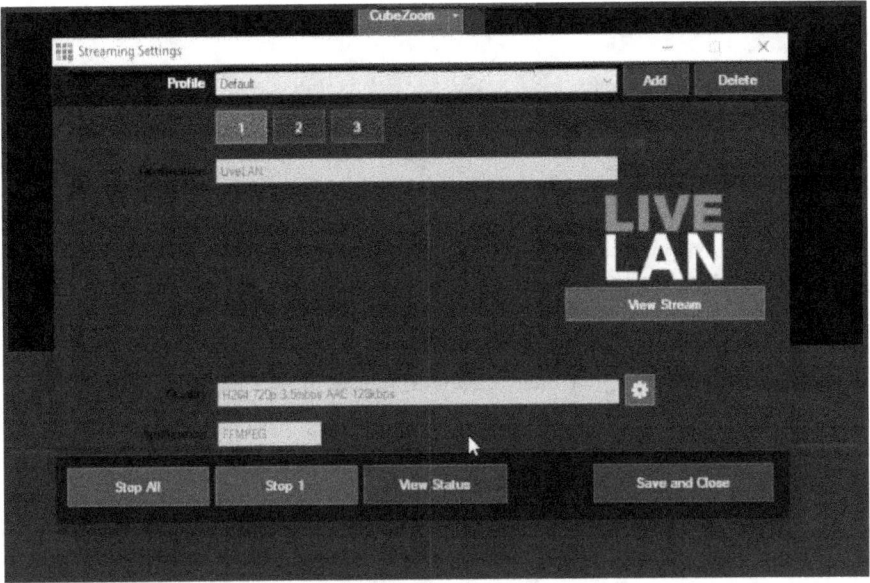

LIVE LAN can be started in the Streaming Settings area of vMix.

LIVE LAN makes it easy for vMix users to stream video to many different types of devices including smartphones, chrome books, and computers such as Windows, Mac or Linux. The wide variety of supported devices makes HLS streaming ideal for organizations using vMix to deliver video to large audiences connected over WiFi or hardwired to the LAN.

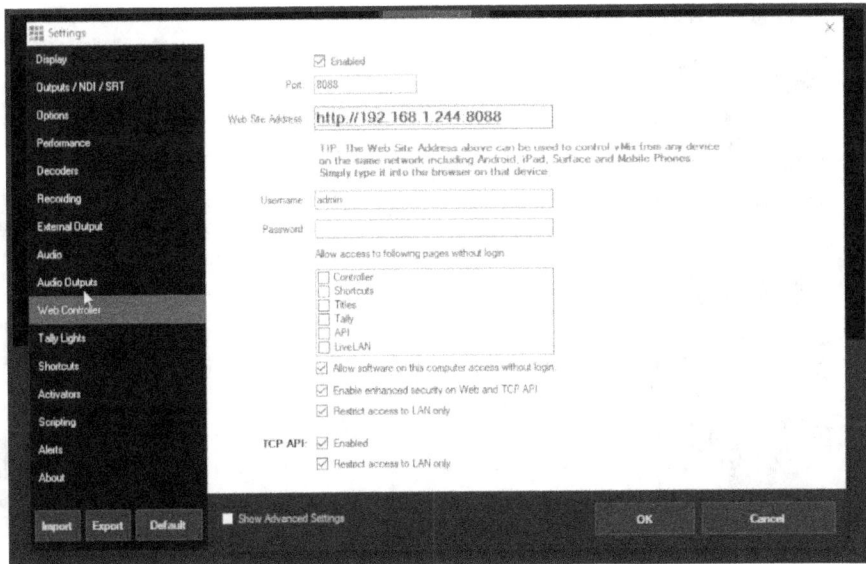

The vMix web controller is used to give or restrict access to LIVE LAN.

To start streaming with Live LAN, vMix has included a "LIVE LAN" streaming destination option which uses the vMix Web Browser tool for users to connect. Therefore, you can visit the Web Browser Settings area in vMix to retrieve the IP address users will need to connect to the live stream. For security purposes, it is suggested that you use a password for important areas of the web browser controller. For example, you would not want to give out the IP address and port number of your vMix LIVE LAN address to students at a school without using a password for the vMix controller.

Pro Tip: It is possible to use a domain name on a local area network to mask your vMix computer's IP address.

Another feature of the HLS encoding type used by LIVE LAN is the stream delivery method. Unlike most streaming solutions which create a consistent stream of bandwidth, HLS delivers video in chunks which can be adaptive to the environment of your network and allow you to handle increased amounts of simultaneous connections. HLS allows you to deliver video via standard web-browsers in an intelligent way. The process of delivering video to modern HTML5 web-browsers via live streaming with HLS was originally designed by developers at Apple. After testing a 3 Mbps 720p video stream set to 30 frames per second, I noticed that the bandwidth fluctuates up and down as the chunks of packets were delivered.

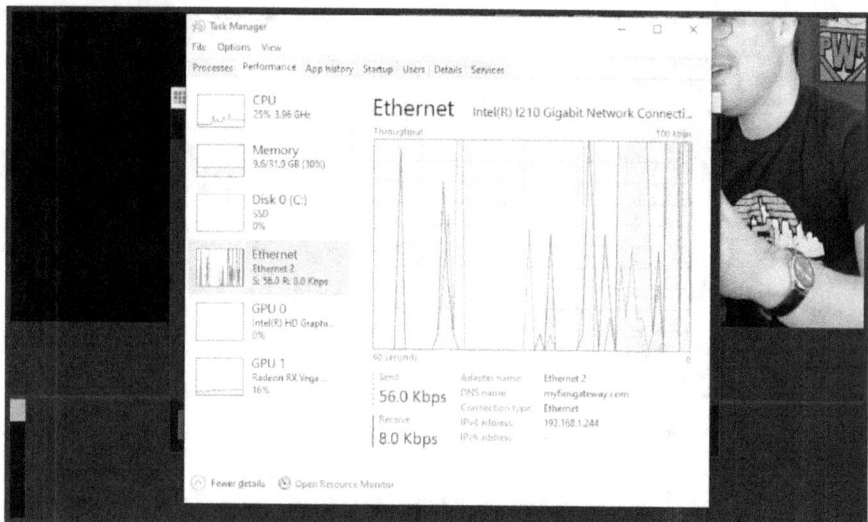

Windows Task Manager can be used to monitor the performance of HLS live streaming in vMix.

vMix API for Developers

vMix is a powerful live video production software platform with functions and features for nearly every need and an intuitive graphical user interface. However, vMix also allows developers to make the software even more powerful by using their API (Application Program Interface.) An API enables developers to create ways to take advantage of the software's functionality from outside of the standard user interface. This creates limitless opportunities for ways of using the software even beyond the ways the original developers intended. APIs can potentially connect vMix to other software, hardware, or interfaces.

Discovering Possibilities

If you are a developer, it may create an interest in developing new applications or tools. For vMix users, it may move them to learn the requisite skills or partner with a developer to bring some new ideas to life.

The Interface in Action

vMix provides two different interfaces for developers, custom XAML, and API, which we refer to here. With the vMix HTTP Web API, many of the software's functions can be controlled by sending a simple command over the network. It is so simple that you can see it in action. Go into your vMix software settings and go to the Web Controller menu. Copy the Website address for the controller. Open a web browser on another computer connected to the same network and enter that address, plus the following: /api/?Function=Fade&Duration=1000.

So, for instance, if my web controller website address was http://192.168.0.128:8088, I would enter http://192.168.0.128:8088/api/?Function=Fade&Duration=1000 into the address bar of my web browser and hit enter. If you entered everything correctly, you should see whatever is in your vMix preview window, move to the output window with a one-second fade.

Available Functions

Most users will not access these commands through a web browser, but the effect is the same. The web API call can access any of the functions available through the Shortcuts feature in vMix. You can see that list by going into the vMix settings menu, selecting Shortcuts, adding a new shortcut, and clicking on the Function field's dropdown menu. They are sorted into categories.

That function needs to be specified in the API call along with any necessary parameters, including duration, input source, text for titles, or other values. The necessary parameters can be found, again, in the add shortcut dialog box in the settings menu. If the function is successfully carried out, it will return a standard response code.

Once you understand the basic framework of the vMix API, all it takes is an understanding of using developing a program, app, or applet to communicate with vMix. Information on that type of programming is available on the web, or you can find a developer with experience with APIs.

AN AMAZING CASE STUDY

How the Griswold High School Uses vMix to Produce the Morning Announcements

The Griswold High School morning announcements show manages to involve over 10 real-time student-run operational roles using vMix and an IP connected workflow. Throughout the production of their morning announcements show, the club is using vMix in multiple ways on separate computers. Leveraging the school's local area network (LAN), the club is able to connect four computers together into a single cohesive video production environment. Each computer provides an individual student access to the main vMix system and it allows students to interact with each other and share resources.

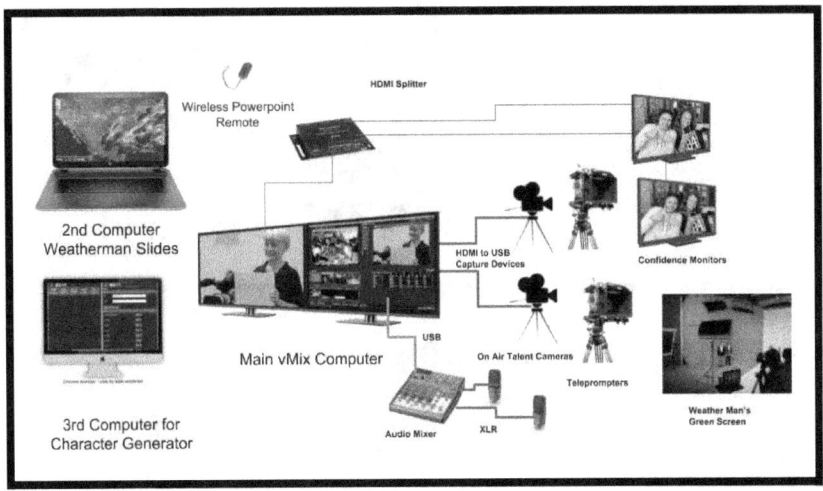

The graphic above illustrates the major parts of the video production system. The main computer is running vMix and is controlled by the technical director using a keyboard and mouse. This computer has three camera sources and the audio mixer connected to it. The second computer is used to display the weathers man's slides for the day. Using a camera with a green screen, the main vMix computer can chroma key the weather man's background and put him in front of his PowerPoint slides. In this way, the school can present the weather just like you see on TV and the weatherman can control his slides with a wireless PowerPoint remote control that is used to advance the slides.

This laptop is running a software called the vMix Desktop Capture which is able to send the video directly from the laptop into the main vMix computer for the technical director to use over the network. You can see a "Confidence" monitors is in place to show the weatherman what he looks like on camera with his PowerPoint slides behind him. This allows the weatherman to see exactly what he is talking about while remaining focused on his delivery to the camera.

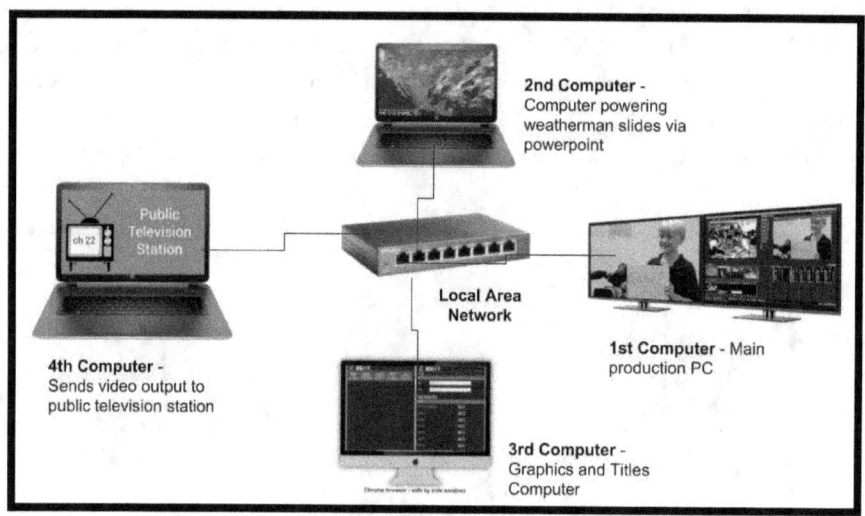

A third computer is used as a character generator. This computer uses the vMix Web Controller to allow the graphics chief to update titles inside vMix and overlay graphics directly onto the broadcast when they are needed using shortcuts.

The vMix web interface has 4 different areas: shortcuts, controller, tally lights, and titles. These can be changed by clicking the icons along the top. For the graphics chief, they are only concerned with titles and shortcuts. The graphics chief can open two google chrome web-browsers and split the screen 50/50 to have access to both at the same time. On the titles screen, the graphics chief will see all the titles that are currently in the vMix production on the main computer. From here the operator can quickly edit and change these titles before they are shown on screen.

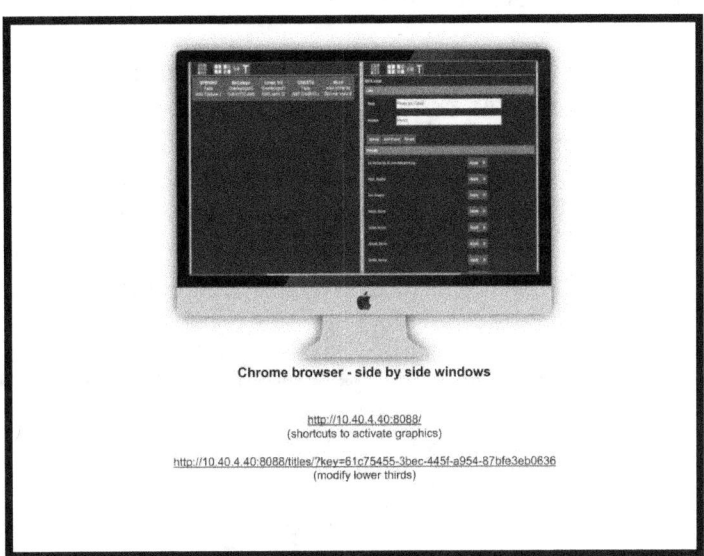

Chrome browser - side by side windows

http://10.40.4.40:8088/
(shortcuts to activate graphics)

http://10.40.4.40:8088/titles/?key=61c75455-3bec-445f-a954-87bfe3eb0636
(modify lower thirds)

The shortcuts screen will automatically show all shortcuts that have been set up on the main vMix machine. A shortcut is a trigger that can be designed to perform almost any video production task in the software. This allows the club to give the graphics chief access to specific buttons that can trigger almost any action including fading to the opening video, overlaying graphics, and initiating the closing credits. In the picture above, you can see the shortcut buttons available on the left-hand side of the screen. These include opening, birthdays, lower third, credits, and word of the day.

The word of the day is an interesting portion of the show where the club leverages website data to produce up to date information each day. The club uses vMix's "Web Browser" input to display this information by entering in the address to the Dictionary.com's word of the day website (https://www.dictionary.com/wordoftheday). Because this webpage is updated every day, the input will always have a new word displayed in its title each day. So, when it comes time to display the word of the day, the graphics chief only needs to click the shortcut button that has been configured to overlay a cropped portion of this webpage onto the screen. The teleprompter chief will check this word and make sure it is included in the script for the on-air talent. The on-air talent will, in turn, make sure they understand the correct pronunciation before the show.

The school birthday title is a ticker that can be updated with information that scrolls across the bottom of the screen. Tickers are used quite commonly in television production to provide additional information in a non-obtrusive way. The graphics chief can quickly update this information inside the titles section of their web controller. If there is a birthday that they need to display in this ticker, they can use the shortcut button to trigger the overlay on and off the broadcast as needed.

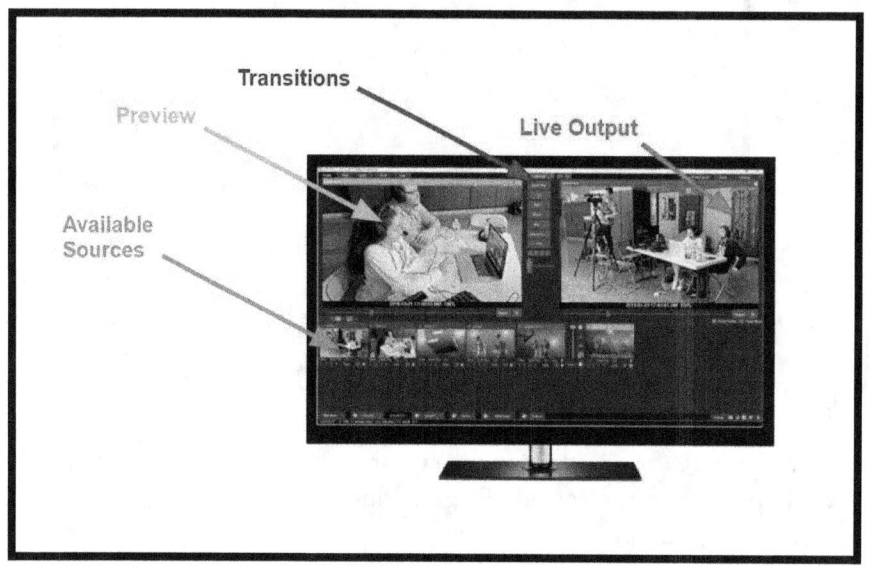

Inside vMix, each camera input looks like a square. Each square has a number which corresponds to its position in the production. When the technical director clicks one of these squares, that input is put into the preview window by default. This preview window is the area that the technical director uses to queue up the next upcoming video input for the production.

Using Multiview layers the technical director can have a single input setup with multiple layers attached by default into a new composition. The broadcast club does this nicely with the weatherman input. This input is essentially two inputs layered together into one. Using this method, the technical director has just one input to transition to that is built out of the weatherman PowerPoint slides with the chroma keyed video layered on top. Other important buttons that you will learn about in our online course include the record button, the stream button, and the fullscreen button.

The technical director has a total of 15 inputs inside vMix. The graphics chief controls five of these inputs and the other 10 have been organized into a production workflow for morning announcements. Here are the inputs:

1-4: Camera placeholders (makes shortcuts easier to manage)
5-7: Camera inputs (HDMI internal card, external dual HDMI capture)
8: Opening graphics (Graphics Chief)
9: Virtual camera input with chroma key enabled
10: Credits (Graphics Chief)
11: Birthday Ticker (Graphics Chief)
12: Lower Third (Graphics Chief)
13: NDI in from Weatherman PowerPoint Slides (Weather)
14: Web Browser - (input from dictionary.com cropped for word of the day) (Graphics Chief)
15: Audio Line In

It is really quite incredible what can be done with this type of network connected video production software. This level of video production quality used to cost hundreds of thousands of dollars and therefore was out of reach of most broadcast clubs. Most traditional video production switchers have a centralized control system. This type of centralized control has traditionally made it difficult to have multiple people working together using standard computers. Now school broadcast clubs can build out systems that can integrate teams of students with independent roles working together on a single broadcast simultaneously.

CONCLUSION

Over the years, vMix has made many significant steps to improve their software for the users they support. I have had the privilege to meet up with the creators each year at the NAB Show (National Association of Broadcasters) in Las Vegas, Nevada. I have seen firsthand that vMix consistently listens to their customers to create a product that is truly designed for modern live video producers. Each major new release has built upon the last adding new breakthrough features such as NDI support, vMix Call, Color Correction, SRT and so much more. While no Macintosh version of vMix exists, for those using Windows computers, you will be hard pressed to find a more robust solution for video production.

A lot can be said about choosing a video production software that you grow into. There is a significant learning curve users face when they start using a new software solution. Hours of time are spent learning a new interface which can be wasted if you eventually outgrow the feature set available on the software you started learning on. Many users who start with the free OBS (Open Broadcaster Software) solution eventually find themselves upgrading to vMix in order to gain the features they need to complete their video production projects.

Luckily, vMix allows users to try the software for free for 60 days. Not only that, but vMix offers tiers of price points that make it easy for organizations to upgrade and find the solution that fits their budget.

Finally, I would like to express how important vMix has been for my own career. As a video production professional, I use vMix more often than Adobe Photoshop, Premiere, or any other primary video software tool I have access to. I even use vMix in combination with Zoom to spice up my meetings and host private webinars where I can switch between multiple cameras with ease.

It has taken many years for me to become proficient in vMix. It is my hope that this book will speed along the learning process for aspiring video professionals around the world.

Cheers,

Paul Richards

Feel free to email me directly at,

Paul.Richards@streamgeeks.us

GLOSSARY OF TERMS

3.5mm Audio Cable - Male to male stereo cable, common in standard audio uses.

4K - A high definition resolution option (3840 x 2160 pixels or 4096 x 2160 pixels)

16:9 [16x9] - Aspect ratio of 9 units of height and 16 units of width. Used to describe standard HDTV, Full HD, non-HD digital television and analog widescreen television.

API [Application Program Interface]- A streaming API is a set of data a social media network uses to transmit on the web in real time. Going live directly from YouTube or Facebook uses their API.

Bandwidth - Bandwidth is measured in bits and the word "bandwidth" is used to describe the maximum data transfer rate.

Bitrate – Bitrates are used to select the data transfer size of your live stream. This is the number of bits per second that can be transmitted along a digital network.

Broadcasting - The distribution of audio or video content to a dispersed audience via any electronic mass communications medium.

Broadcast Frame Rates - Used to describe how many frames per second are captured in broadcasting. Common frame rates in broadcast include **29.97fps and 59.97 fps**.

Capture Card - A device with inputs and outputs that allow a camera to connect to a computer.

Chroma Key - A video effect that allows you to layer images and manipulate color hues [i.e. green screen] to make a subject transparent.

Cloud Based-Streaming - Streaming and video production interaction that occurs within the cloud, therefore accessible beyond a single user's computer device.

Color Matching - The process of managing color and lighting settings on multiple cameras to match their appearance.

Community Strategy - The strategy of building one's brand and product recognition by building meaningful relationships with an audience, partner, and clientele base.

Content Delivery Network [CDN] - A network of servers that deliver web-based content to an end user.

CPU [Central Processing Unit] – This is the main processor inside of your computer, and it is used to run the operating system and your live streaming software.

DAW - Digital Audio Workstation software is used to produce music. It can also be used to interface with multiple devices and other software using MIDI.

DB9 Cable - A common cable connection for camera joystick serial control.

DHCP [Dynamic Host Configuration Protocol] Router - A router with a network management protocol that dynamically sets IP addresses, so the server can communicate with its sources.

Encoder - A device or software that converts your video sources into an RTMP stream. The RTMP stream can be delivered to CDNs such as Facebook or YouTube.

FOH – Front of House is the part of your church that is open to the public. There is generally a FOH audio mix made to fill this space with the appropriate audio.

GPU – Graphics Processing Unit. This is your graphics card which is used for handling video inside your computer.

H.264 & H.265 - Common formats of video recording, compression, and delivery.

HDMI [High Definition Multimedia Interface] - A cable commonly used for transmitting audio/video.

HEVC [High Efficiency Video Coding] - H.265, is an advanced version of h.264 which promises higher efficiency but lacks the general support of h.264 among most software and hardware solutions available today.

IP [Internet Protocol] Camera/Video - A camera or video source that can send and receive information via a network & internet.

IP Control - The ability to control/connect a camera or device via a network or internet.

ISP – Internet Service Provider. This is the company that you pay monthly for your internet service. They will provide you with your internet connection and router.

Latency - The time it takes between sending a signal and the recipient receiving it.

Live Streaming - The process of sending and receiving audio and or video over the internet.

LAN [Local Area Network] - A network of computers linked together in one location.

MIDI [Musical Instrument Digital Interface] - A way to connect a sound or action to a device. (i.e. a keyboard or controller to trigger an action or sound on a stream

Multicast - Multicast is a method of sending data to multiple computers on your LAN without incurring additional bandwidth for each receiver. Multicast is very different from Unicast which is a data transport method that opens a unique stream of data between each sender and receiver. Multicast allows you to broadcast video from a single camera or live streaming computer to multiple destinations inside your church without adding the bandwidth burden on your network.

Multicorder – Also known as an "IsoCorder" is a feature of streaming software that allows the user to record raw footage from camera feed directly to your hard drive. This feature allows you to record multiple video sources at the same time.

NDI® [Network Device Interface] - Software standard developed by NewTek to enable video-compatible products to communicate, deliver, and receive broadcast quality video in high quality, low latency manner that is frame-accurate and suitable for switching in a live production environment.

NDI® Camera - A camera that allows you to send and receive video over your LAN using NDI technology.

NDI®|HX - NDI High Efficiency, optimizes NDI for limited bandwidth environments.

Network - A digital telecommunications network which allows nodes to share resources. In computer networks, computing devices exchange data with each other using connections between nodes.

Network Switch – A network switch is a networking device that connects multiple devices on a computer network using packet switching to receive, process and forward data to the destination device.

NTSC - Video standard used in North America.

OBS – Open Broadcaster Software is one of the industries most popular live streaming software solutions because it is completely free. OBS is available for Mac, PC, and Linux computers.

PAL - Analog video format commonly used outside of North America.

PCIe- Allows for high bandwidth communication between a device and the computer's motherboard. A PCIe card can installed inside a custom-built computer to provide multiple video inputs (such as HDMI or SDI).

PoE - Power over Ethernet.

PTZ - Pan, tilt, zoom.

RS-232 - Serial camera control transmission.

Router – Your internet router is generally provided to you by your internet service provider. This device may include a firewall, WiFi and/or network switch functionality. This device connects your network to the internet.

RTMP [Real Time Messaging Protocol] – Used for live streaming your video over the public internet.

RTSP [Real Time Streaming Protocol] - Network control protocol for streaming from one point to point. Generally, used for transporting video inside your local area network.

vMix® – vMix is a live streaming software built for Windows computers. It is a professional favorite with high-end features such as low latency capture, NDI support, instant replay, multi-view and much more.

Wirecast® – Wirecast is a live streaming software available for both Mac and PCs with advanced features such as five layers of overlays, lower thirds, virtual sets and much more.

xSplit® – xSplit is a live streaming software with a free and/or low monthly fee paid option. This is a great software available on for Windows computers that combines advanced features and simple to use interface.

ABOUT THE AUTHOR

Paul is the Chief Streaming Officer for StreamGeeks. StreamGeeks is a group of video production experts dedicated to helping organizations discover the power of live streaming.

Every Monday, Paul and his team produce a live show in their downtown West Chester, Pennsylvania studio location. Having produced live shows as amateurs themselves, the StreamGeeks steadily worked their way to a professional level by learning from experience as they went.

Today, they have an impressive following and a tight-knit online community which they serve through consultations and live shows that continue to inspire, motivate, and inform organizations who refuse to settle for mediocrity. The show explores the ever-evolving broadcast and live streaming market while engaging a dedicated live audience.

As a husband and father raising his children in the Lutheran faith, Richards knows a thing or two about the technology inside the church. Richards now specializes in the live streaming media industry leveraging the technology for lead generation. In his book, "Live Streaming is Smart Marketing", Richards reveals his view on lead generation and social media.

Additional Online Courses:

Join over 20,000 other students learning how to leverage the power of live streaming! Take the following courses taught by Paul Richards for free by downloading the course coupon codes available at streamgeeks.us/start.

- **Facebook Live Streaming** - *Beginner*

This course will take your through the Facebook Live basics. It has already been updated twice! This also includes using Facebook Live Reactions!

- **YouTube Live Streaming** - *Beginner*

This course will take you through the YouTube Live basics. It also includes essential branding and tips for marketing.

- **Introduction to OBS (Open Broadcaster Software)**

This course will take your through one of the world's most popular FREE live streaming software solutions. OBS is a great place to start live streaming for free!

- **Introduction to xSplit Software** - *Beginner*

This course takes you through xSplit which has more features that OBS but costs roughly $5/month. Learn how to create amazing live productions and make videos much faster with xSplit!

- **Introduction to vMix** - *Intermediate*

vMix will have you live streaming like the Pros in no time. This Windows based software will amaze even the most advanced video producers!

- **Introduction to Wirecast** - *Intermediate*

Wirecast is the preferred software for so many professional live streamers. Available for Mac or PC this is the ideal software for anyone looking for professional streaming.

- **Introduction to NewTek NDI** - *Intermediate*

NewTek's innovative IP video standard NDI (Network Device Interface) will change the way you think about live video production. Learn how to use this innovative new technology for live streaming and video production system design.

- **Introduction to live streaming course** - *Beginner*
This course includes everything you need to get started designing your show. This course includes a starter pack of course files including: Photoshop, After Effects and free Virtual Sets.

- **Introduction to live streaming** - *Intermediate*
This course focuses on more advanced techniques for optimizing your production workflow and using compression to get the most out of your processor. This course includes files for: Photoshop, After Effects and free Virtual Sets.

- **Live Streaming for Good - Church Streaming Course** - *Intermediate*
This course focuses on live streaming for churches and houses of worship. We tackle some of the big questions about live streaming in a house of worship and dive into the specific challenges of this space.

- **How to Live Streaming A Wedding** - *Beginner*
This is a great course for anyone looking to start live streaming weddings. Originally designed for Wedding Photographers to add a live streaming service to their existing portfolio of offerings. This course is great for beginner